QUEEN ELIZABETH HOUSE

Modern Processing, Packaging and Distribution Systems for Food

Edited by

FRANK A. PAINE, B.Sc., C.Chem., F.R.S.C., F.I.F.S.T., F. Inst. Pkg
Secretary-General
International Association of Packaging Research Institutes
and
Adjunct Professor, School of Packaging
Michigan State University

Blackie
Glasgow and London

Published in the USA by
avi, an imprint of
Van Nostrand Reinhold Company
New York

Blackie and Son Limited
Bishopbriggs, Glasgow G64 2NZ
7 Leicester Place, London WC2H 7BP

Published in the USA and Canada by
AVI, an imprint of
Van Nostrand Reinhold Company Inc.
115 Fifth Avenue
New York, New York 10003

Distributed in Canada by
Macmillan of Canada
Division of Canada Publishing Corporation
164 Commander Boulevard
Agincourt, Ontario M1S 3C7

© 1987 Blackie and Son Ltd
First published 1987

British Library Cataloguing in Publication Data
Modern processing, packaging and distribution
systems for food.
1. Food industry and trade
I. Paine, F.A.
664   TP370

ISBN 0-216-92247-X

For the USA and Canada

ISBN 0-442-20510-4

Photosetting by Best-set Typesetter Ltd
Printed in Great Britain by Bell & Bain (Glasgow) Ltd

Preface

The progress that has been made over the last decade in the preparation, development, processing, and marketing of food has to a large extent been made possible by innovations and developments in the ways that thermoplastics, in conjunction with paper, metal foils, adhesives and other materials, have been combined and formed into the appropriate configurations to provide the properties required.

Much has been said, written and published about retort pouches, modified atmosphere packaging and aseptic preservation processes, and even more about the newer methods of distribution and retailing of all kinds of food. However, all of this material needed to be digested, condensed into a logical framework and appraised, and possible further developments considered.

In many instances, the original research and development was carried out in conjunction with one or more of the research organisations in membership with IAPRI, the International Association of Packaging Research Institutes, and it was felt that a book which attempted to provide a review of the more important developments would be useful to practitioner and student alike. This volume therefore aims to provide the food technologist with a more comprehensive understanding of the packaging aspects of the marketing and distribution of food; the packaging technologist with an insight into the basics of food technology that must influence his thinking; and the business executives concerned with food and its distribution with a readable account of past achievements and possible future developments. The student and industrial trainee should also benefit from a study of the ways in which these developments have taken place.

Each of the chapters was written by a senior member of one of the leading IAPRI Institutes which has made a study of the subject matter and can consider the future possibilities with the least commercial bias. As it was thought desirable that each chapter should provide a complete picture of its area, there is inevitably a small overlap between certain chapters. However, this has been kept to a minimum by cross-referencing.

My personal thanks are due to all of the authors for improving the format of their chapters from the layout that I originally suggested, and for accepting the relatively small amount of editing that was needed to bring the work into what, I trust, is an integrated account of the more recent developments in food packaging.

FAP

Contents

3 Modified atmosphere packaging
RICHARD INNS

4 Use of irradiation techniques in food packaging
KIRSTEN NIELSEN

5 Shelf-life prediction
DENNIS J. HINE

9 Packaging for consumer convenience
INGRID FLORY

List of contributors

Ivanka Dimitrova   Project Leader, The Swedish Packaging Research Institute, Torshamnsgatan 24, Box 9, S-163 93 Spanga, Stockholm, Sweden.

Ingrid Flory   Consultant and Project Leader, c/o The Swedish Packaging Research Institute, Torhamnsgatan 24, Box 9, S-163 93 Spanga, Stockholm, Sweden.

Roger Griffin   Visiting Lecturer, School of Packaging, Michigan State University. East Lansing, Michigan 48824-1223, USA.

Dennis Hine   Principal Consultant, Packaging Division, Pira, The Research Association for Packaging, Leatherhead, Surrey, KT22 7RU, UK.

Richard Inns   Director, Packaging Division. Pira, The Research Association for Packaging, Leatherhead, Surrey KT22 7RU, UK.

Arve Iversen   Project Leader, Norwegian Food Research Institute, PO Box 50, N-1432 Aas-NLH, Norway

Loa Karjalainen   Director, Association of Packaging Technology and Research, c/o Finnish Pulp and Paper Research Institute, Box 136, SF-00101 Helsinki, Finland.

Kirsten Nielsen   Manager, Packaging Materials Section, Danish Packaging, Distribution and Transport Research Institute, Meterbuen 15, DK-Skovlunde, Denmark.

Frank Paine   Secretary General, International Association of Packaging Research Institutes, Eyot Lodge, Petworth Road, Chiddingfold, Surrey, GU8 4UA, UK.

1 Retortable plastic packaging

ROGER C. GRIFFIN

The origin of retorting as a food process

The processing of foods to make them easier to eat and easier to store is as old as archaeological records. Smoking, salting, sun-drying, and fermentation techniques all ante-date historical records. Nevertheless, until 200 years ago, civilized mankind had only these very limited means for preserving perishable foods beyond harvest time. The period between harvests could and often did become a time of fasting, rationing, or even famine for most of the population. Many of the sauces and spices used in cooking were developed to mask the flavours of tainted meats. How many people died of eating contaminated foods can only be conjectured, and vitamin deficiency diseases were rampant, particularly scurvy.

All of these limitations were multiplied when it came to military operations. Armies were disbanded or confined to winter quarters until adequate supplies of food could be obtained to sustain a campaign. Armies surrendered for lack of food or retreated when decimated by disease or hunger. Fortresses fell more often from starvation than by storm, and it was Napoleon Bonaparte who recognized that an army travelled on its stomach and offered a great monetary prize to anyone who would find a better way to preserve food.

Nicholas Appert claimed the prize and is credited the inventor of heat-processing, from which retorting evolved. Appert cleaned wine bottles, stuffed food such as beef and broth into the bottles, placed the bottles in boiling water for several hours and then sealed them with tightly-fitted corks. The resulting foods were overcooked, but mostly did not spoil when held at room temperature. Appert had no idea of the technology of food spoilage; the work of Koch and Pasteur came more than a half-century later, and heat-processing of food was a trial and error technique for nearly 100 years.

Appert's pioneering efforts were improved upon. Peter Durand developed the soldered metal can in 1810. Michael Faraday suggested using brines to raise the temperature of the processing tanks. In 1874 the pressurized steam-heated retort was invented by A.K. Shriver, and the 'sanitary' double-seamed three-piece can was introduced by Max Ams in 1888.

Pressurized retorts have been improved upon continuously up to the present day. There are now batch-type and continuous-type, agitating and still, horizontal and vertical, steam, steam/air, and steam/water/air processes. They can be automated, controlled, monitored, and instrumented. Temperatures as high as 150°C and process times of a few seconds are now more than just experimental.

Thermal processing of foods in retorts remained an art rather than a science until 1920–30, when scientists and engineers, mathematicians and microbiologists turned their attention to the problem. As a result of their efforts a technology has been developed which has increased speeds of filling and closing, shortened process times, improved quality, reduced costs, and raised the safety and dependability of retort food processing to levels undreamed of in Appert's day. Billions of cans of food are processed annually, and the number of deaths attributable to food poisoning from improper processing has been reduced to single-digit figures.

Traditional packaging

Metal cans

The three-piece can has been the work-horse of the retorted food processing industry. Traditionally it was made from tin-plated steel with interior lacquers to prevent interaction with foodstuffs. Body seams which were hooked, bumped, tacked and soldered are now being increasingly supplanted by welded seams.

Because of the high cost of tin, the weight of tin-plating has been reduced gradually, concurrent with techniques for achieving greater uniformity of thickness.

Aluminium made its first inroads in the can market as a shallow drawn two-piece can for foods such as tuna fish and vienna sausage. Its major penetration came with the development of the two-piece beverage can made by draw–redraw or by draw-and-iron processes. As manufacturing experience led to refined techniques, the wall thicknesses were successively reduced to the point that the pressure of the carbonation was needed to prevent compression crushing (for non-carbonated beverages the pressure is provided by adding a measured quantity of liquid nitrogen just before closing). To date, the aluminium can has not made major inroads in retortable food markets.

Glass jars

With the development of suitable metal cap closures, glass jars made substantial penetration of the retorted food markets. Their major niche was in acid foods such as fruits and pickles, acidified low-acid foods such as onions, and low-acid foods where display of the product was an aesthetic advantage, as with whole green beans and asparagus stalks.

Baby foods have been dominated by glass jars for many years for several reasons, reclosability and ease of product removal being major advantages.

To achieve the strength required to resist breaking under crushing loads, particularly at the angled top and bottom shoulders, glass containers originally had to be thick-walled throughout. With the development of more sophisticated blow-moulding equipment it is now possible to put thickness where needed and to lighten the weight of the container. Coatings and surface treatments have been developed that reduce contact scratching and the damage that results from the weakened scratched surfaces.

The greatest hazards for glass containers are mechanical or thermal shock. Mechanical shock can occur from impact with a hard surface, such as another container, a machine part or a concrete floor, or it can occur from hydrodynamic impacts which are created in a water-filled retort when steam is introduced too rapidly. Thermal shock occurs when the temperature of the internal surface differs too greatly from that of the exterior surface. Because glass is a poor conductor of heat, the colder surface cannot expand rapidly enough to match the expansion of the hotter surface. This puts the colder surface under tension. Because glass is far weaker under tension than under compression, this stress causes fracture. To avoid thermal and/or hydro-dynamic shock, heating and cooking of foods in glass containers must be slower.

Closures on glass jars are either snap-on or screw-on. For maintenance of an adequate seal throughout the retort process, the cap is usually made of steel. Steel has the strength and rigidity to apply the force which compresses the soft liner against the mouth of the jar. When contents are filled hot, a partial vacuum is formed on cooling, which establishes the initial seal. Screw-on caps are retorqued after cooling as a precaution against loosening.

Advantages and disadvantages

Metal cans and glass jars have monopolized the market for retorted foods because there were no economical substitutes. Each possesses both advantages and disadvantages (Table 1.1).

Table 1.1 Properties of metal cans and glass jars

Property	Metal cans	Glass jars
Barrier	Excellent	Excellent (except to light)
Product visibility	None	Good
Weight	Lighter	Heavier
Production speeds	Faster	Not as fast
Heat transfer	Faster	Slower
Strength (crush)	Good	Better
Ease of opening	Not as good	Excellent
Reclosability	Poor	Excellent
Breakage/damage resistance	Fair to good	Not as good
Require lacquering	Yes	No

Thermal processing

Retorting, commonly called 'canning', is a form of thermal processing of foods. A brief review of the basic principles of retorting will help clarify the advantages of packages that will be discussed in the following sections.

Spoilage organisms

Foods are spoiled principally by the action of micro-organisms. These are living creatures that multiply, mature, and die. During their life cycle they 'eat' nutrients and 'excrete' waste products. These products can be beneficial or harmful. Examples of beneficial actions are conversion of sugars by yeasts to produce alcohol and carbon dioxide from which beer, wine, and spirits can be made; conversion of apple juice to vinegar; conversion of cabbage to sauerkraut; ripening of cheeses, and manufacture of yoghurt. Examples of harmful actions are sliming, moulding, souring of meats and vegetables, and rotting and moulding of fruits.

There are some species of micro-organisms that cause spoilage that can be toxic or fatal to humans if they eat the spoiled food. There are two basic types; those that produce toxins in the food which are the harmful agents, and those which merely infest the food and then grow within the body to produce illness or death. Examples of toxin producers are shown in Table 1.2.

Table 1.2 Spoilage organisms

Toxin-producing organisms	Toxin-induced illness
Clostridium botulinum	Botulism
Claviceps purpurea	Ergotism
Aspergillus flavus	Aflatoxin poisoning
Infesting organisms	Illness
Salmonella (various)	Severe intestinal
Shigella dysenteria	distress, vomiting, etc.,
Staphylococcus aureus	e.g. 'Montezuma's Revenge'
Escherichia coli	

Each species and variety of micro-organism requires favourable conditions for optimal growth. These conditions vary for each organism, and broad classifications are made relating to these conditions.

Factors influencing growth

Environmental factors that influence the growth of micro-organisms include temperature, available moisture, acidity, presence or absence of oxygen, and availability of nutrients. Conversely, growth may be inhibited by accumulation of waste products and lack of one or more of the favourable growth factors. For example, spoilage can be inhibited by refrigeration and essentially halted by freezing. Likewise, removal or capture of available water by drying or by adding sugar or salt will inhibit growth.

Storage of moist, low-acid foods at room temperature presents ideal conditions for growth of many spoilage organisms, including, when oxygen is excluded, the very dangerous *Clostridium botulinum*.

Thermal processing is undertaken to destroy by heat the micro-organisms that cause harmful food spoilage when food is held at room temperature. Some organisms (yeasts and moulds) are killed at temperatures below the boiling point of water. Others are very heat-resistant, even to the point of forming dormant spores under adverse conditions. These spores do not multiply but can survive extreme conditions. When the environment again becomes favourable for growth, the spores regenerate to normal viable cells. To kill spore-forming organisms, temperatures well in excess of that of boiling water must be employed.

The aim of thermal processing is to bring food as rapidly as possible to a temperature at which unwanted organisms are killed and to hold the food at that temperature long enough to render the food 'commercially sterile.' Commercial sterility is achieved when all life-threatening and hazardous-to-health organisms (pathogens) are killed, and organisms capable of inducing other spoilage of the food are reduced in number to no more than one in 10 000 cans.

D value. While, on average, a population of a particular micro-organism has a sensitivity to heat, individual sensitivities will vary. It can be established by careful measurement, using pure strains of the organism and controlled conditions, how much time at a specified temperature will be required to kill 90% of the organisms present. This is called the thermal death time (TDT). Or, since it represents a decimal reduction i.e. 10% survivors, it is also called the D value. It has been demonstrated that a second interval of D minutes at the same temperature will kill 90% of the survivors i.e. an additional decimal reduction. Now only 1% of the original population survives.

If the food could be raised instantaneously to the desired process temperature and held there for $5D$ minutes, 0.001% of the original population of organisms would survive.

Heat penetration. In practice, food cannot be instantaneously heated. The container is placed in a retort and the retort is brought up to process temperature as rapidly as possible. The outer layers of the food are heated first and this heat is then transferred to inner layers by conduction or convection. The colder the food to start with, and the greater the distance the heat must penetrate, the longer it will take to bring the coldest area up to process temperature. Heat penetration rates need to be established before a dependable process time can be calculated. However, there are a few more factors to be considered.

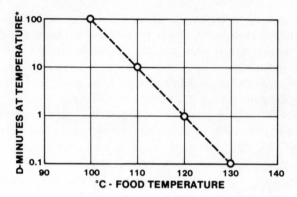

Figure 1.1 D values for $Z = 10°C$. D = minutes at temperature to destroy 90% of organisms present.

Z value. It has been shown that D values are smaller at higher temperatures. A plot of D values (time) v temperature on semi-logarithm scale will result in a straight line. The number of degrees (Celsius) increase in food temperature required to shorten the D value by a factor of 10 is called the Z value of the organism in the food medium tested. Figure 1.1 illustrates a plot of D values for an organism with $Z = 10°C$. By means of this relationship an equivalent killing power can be calculated for any temperature.

In Figure 1.1, for example, 10 minutes at 110°C achieves the same decimal reduction of organisms as does 1 minute at 120°C or 0.1 minute at 130°C. Likewise, 1 minute at 110°C is equivalent in lethality to 0.1 minute at 120°C, etc.

F_0 value. The F_0 value of a process is the number of minutes at 250°F (121°C) to which the coldest part of the food must be subjected in order to achieve the desired commercial sterility.

Heat penetration tests determine how many minutes the food remains at each temperature increment. These actual minutes are converted to equivalent minutes at 250°C (121°C) by tables of D *and* Z values. The summation of the equivalent minutes is the actual lethality of the process.

The targeted or required F_0 of the process must be selected from prior knowledge of trained experts, and must take into consideration probable bacterial load. Where prior experience is lacking, then inoculated test packs are mandatory.

Bacterial load. To calculate the number of D values (decimal reductions) needed to achieve commercial sterility (not more than 1 viable spore in 10 000 containers) it is necessary to have a reasonable estimate of the bacterial load, i.e. the population of bacteria in each container at the start of the retort process. There are microbiological laboratory techniques for estimating this, but much depends upon factors that are difficult to control. These include plant sanitation, product condition at the start of pre-processing, length of time product is held prior to retorting, temperature of holding, and slowness of retort coming to process temperature.

The monitoring of processes by the US industry-sponsored National Food Processors' Association (formerly National Canners Association) has made it possible in most instances to make educated estimates of required F_0 values for many foods in conventional containers based on their recommendations.

Inoculated test packs. Inoculated test packs consist of a series of retort processes on batches of containers that have been inoculated with a known number of test spoilage organisms. The organisms are deliberately selected to be much harder to kill than the dangerous *Clostridium botulinum*. The series of processes are varied in time to deliberately underprocess and overprocess the batches. By incubating the containers and determining the number of spoiled containers in each batch, it can be determined how long a process (F_0) is needed to achieve no spoilers. This also provides a means for verifying the efficacy of calculated process times by way of heat penetration data.

Disadvantages of can and jar geometry

Penetration of heat depends upon heat transfer from the heating medium through the package wall and hence through half the thickness of the food. In the absence of mixing by convection or agitation, the greater the distance to the centre of the food, then the greater the time lag for heat penetration. In fact it is roughly proportional to the square of the distance, that is, twice the thickness means four times slower for penetration of heat to the centre.

Penetration of heat is also proportionate to the difference in temperature between the heat source and the food. The greater this difference; the faster the heat transfer.

Cans and jars are traditionally cylindrical and typically 3″ (76mm) or more in diameter. With this geometry, outer layers of the food reach process temperature first and are 'cooked' longer than the centre portion. By the time the centre is sterile the outer layers have been overprocessed to the degree that colour, flavour, texture, and/or nutrients are adversely affected. This is most pronounced in viscous or solid foods and least pronounced in low-viscosity liquids where convection provides a mixing of outer and inner 'layers.' Rotation of packages can promote mixing and thereby shorten required process time. Obviously, the greater the distance heat must penetrate, the more the outer layers will be degraded.

There have been two major approaches to overcoming this undesirable overprocessing: aseptic processing (see Chapter 2) and the retort pouch. In aseptic processing, the food is sterilized *before* being placed in the package. This can be done with high efficiency in heat exchanges where the food is pumpable. With the retort pouch, food is sterilized *after* being sealed in the package. The much thinner cross-section of food reduces the time of heat penetration as compared to the can or jar of same capacity.

The retort pouch

In a confirmation of the adage 'history repeats itself', the retort pouch was developed in response to a military need. The US Army had for some time been dissatisfied with the tin canned food in the C-ration. Serious criticisms were related to the injuries inflicted on combat troops when they fell upon the cans tucked into battledress pockets. Discarded tins were utilized by enemy troops in jungle warfare as makeshift hand grendades, land mines, or booby traps. Other injuries were incurred from jagged edges of opened cans. Lesser criticisms involved the difficulty of opening and reheating contents under combat conditions. Other desirable objectives were improvement of food quality and nutrition, reduction in weight, and improvement in efficiency of cubic volume.

Under the sponsorship of the Chicago US. Army Quartermaster Corps (later the Natick Laboratories) first efforts in the mid- to late 1950s were directed toward assessing available materials, package forms, and equipment to determine potential feasibilities. The rectangular fin-sealed pouch was selected as the most promising package form, notwithstanding the lack of dependable retortable materials and the need to develop filling, closing, and processing equipment. Development efforts were directed at all phases of the problem concurrently.

Materials

Availability of plastic films capable of surviving retort conditions was extremely limited in the early 1950s. Polyvinyl copolymer and nylon films are too susceptible to moisture absorption which interfered with the making of good seals. Low- and medium-density polyethylene films are too weak at retort temperatures. High-density polyethylene and polypropylene homo- and copolymers became available almost at the same time. The former, modified with isobutylene rubber, was used as an inner food-contacting heat-seal film by one of the leading laminate developers, but in the long run, cast polypropylene became the material of choice.

The outer film of the composite lamination was needed to provide strength and flex resistance. It needed to be resistant to heat-seal temperatures, printable, and able to withstand retort temperatures without bursting, shrinking, or delaminating. The material of choice which became universally accepted was polyethylene glycol terephalate, sold under the trade name 'Mylar' (E.I. duPont de Nemours, Inc.). Other vendors produced competitive films made from identical or nearly identical polyester resins.

To achieve the level of barrier properties deemed necessary for the 1–2 year shelf life specified by military and/or US commercial markets, it was accepted that an aluminium foil layer was required. Research effort centred around the optimum gauge for strength, barrier, and resistance to flex-cracking. Some manufacturers went to the extreme of incorporating two foil plies in the lamination. It should be noted that the Japanese market accepted pouches made without foil. This was possible because a much shorter shelf life was acceptable in their domestic market, 3–6 months for some products.

An intense development effort on materials centred around the formulation of adhesives and primers capable of laminating the dissimilar plies together with adequate strength to survive all the rigours of heat sealing, retorting, and distribution handling while being acceptable to regulatory agencies from the standpoint of not contributing harmful components to the food.

While the exact formulation is closely guarded, the work-horse adhesive for retortable foil/plastic laminates is manufactured by Morton Chemical Div. of Morton Thiokol, Inc. (Adcote 506–40/Catalyst 9 L10). When properly applied and solvent removed, this two-part adhesive system is 'approved' by the FDA for use in retort pouches and container-lidding stock. The adhesive requires a hot nip 250–270°F (121–132°C) and a post-cure at 120°F (49°C) for 5–7 days to develop the required bond strength.

The same manufacturer has developed a dispersion of amorphous polypropylene in solvent for application to aluminium or steel as an adhesion promoter. In that form it can be used as a prime coat on metal in conjunction with the Adcote adhesive. It has also found use as a bonding tie layer for extrusion coatings of polypropylene on metal. Morprime requires fusing of its particles at 380–400°F (193–204°C).

Applied to the flanges of a metal tray, Morprime can be used to seal lid stock with a destruct bond strength or, with modification, a peelable seal.

A typical retort pouch laminate from outside in is as shown in Table 1.3. Note that printing can be buried between the polyester film and the foil.

Table 1.3 Construction of a typical retort pouch laminate

Thickness	LBS/REAM**	Material
0.5 mil (12μm)	10.8	Polyester film
	.8	Ink
0.1 mil (2.5μm)	2.5	Adhesive
	0.1	Morprime (optional)
0.7 mil (18μm)	30.0	Aluminium foil
	0.1	Morprime (optional)
0.12 mil (3μm)	3.0	Adhesive
3.0 mil (75μm)	38.7	Polypropylene film

 * 1 mil = 0.001 in = 25 microns (μm)
** Lbs/3000 sq. ft.

Pouch fabrication

Pouches can be prefabricated and filled and closed on pouch-fed machinery, or they can be made from roll stock in line with filling and closing. While greater speeds may be achievable from the latter, there is much to be said for using pouch-fed machinery.

The manufacturer of the pouches is responsible for quality of bottom and side seals, and can inspect out poor quality.

Control of temperature and dwell time and uniform application of pressure are absolute essentials for the making of strong seals. This is even more important for the final seal, which has the additional requirement of avoiding contamination of the top seal by food components. Here the use of dripless, positive cut-off fillers and the use of protective shields is mandated. Intermittent-motion machines offer the further hazard of product splashing, which is accentuated by higher speeds.

For all of these reasons pouch-fed equipment is most widely used at this time. Speeds of more than 30 units/min are rarely achieved, which means as many as five filling and closing machines are required to achieve 150 units/min output.

Filling, closing, and air evacuation

Because the quantity of filled product determines the thickness of a filled pouch and the thickness dictates process time, it is imperative to closely control product fill. Overfilling can contribute to splashing and contamination of seal areas and put unnecessary strain on seals during retorting.

Because entrapped air would expand during retorting, overriding air is introduced into the retort as counter-pressure. This air reduces the heat transfer rate of saturated steam. Air within the pouch also interferes with heat transfer. To minimize both of these effects, excess air is eliminated just prior to closing by one of several methods.

(i)   Mechanical squeezing of the pouch raises the liquid level and displaces air (it also risks seal area contamination)
(ii) Hot-filling expands the headspace air, expelling the excess, thus reducing its pressure and ultimate volume
(iii) Steam flushing is accomplished by injecting clean superheated steam into the pouch just prior to closing. Air is both diluted and flushed out, while superheating prevents condensation of water in the seal area. Pouches exhibit considerable vacuum on cooling; volumes of entrapped air as little as 1 ml per 200 g of contents are consistently achievable
(iv) Vacuum-sealing requires special vacuum chambers which pump out air prior to making the top seal. Upon release of the vacuum from the chamber, the pouch walls collapse tightly around the contents. This is a

slower operation because of the time required to pump down the chamber. The filled product cannot be too hot, or liquids will flash boil at reduced pressure.

Retorting

There are a number of requirements in the retorting of pouches that differ from typical canning procedures:

(i) Because process times are shorter, precise and preferably automatic control of temperatures, pressures, and process timing are indicated
(ii) In most instances overriding air pressure is required to prevent over-expansion of pouches during heat processing
(iii) In all instances overriding air pressure is essential during cooling to prevent pouches from exploding
(iv) Because filled pouch thickness is a critical factor in determining process time, pouches must be contained in special racks that prevent overlapping and maintain horizontal orientation
(v) Racks must be designed to assure free flow of heating medium around every pouch without causing pouches to be buffetted about — this is particularly important in water-cook processes
(vi) Pouches should be cooled to well below 100°C prior to release of pressure on opening of the retort
(vii) Pouches should be dried to prevent staining — this can be accomplished by high-velocity air jets.

Testing requirements

The testing of retort pouches is aimed at six critical areas:

(i) Is the laminate properly made (and able to withstand retorting)?
(ii) Are the three manufacturer's seals properly made?
(iii) Is the final closure properly made?
(iv) Is the product fill and residual air content as specified?
(v) Are there any evidences of damage or incipient failure that could lead to post-processing recontamination?
(vi) Will pouches survive laboratory abuse tests and in-use shipping tests?

Laminate testing. Tests on the laminated material are conducted by the converter and checked by the pouch fabricator (if not the same company). These are principally physical, to assure the laminate was made properly from the correct compenents, and may include tests on thickness, yield (sq. in/lb), tensile strength (lbs/in width), % elongation at failure, Mullen burst (gauge psi), and bond strength (peel) (lb/in width). There are two schools of thought on bond strength: those that feel the higher the better, and those that feel a minimum strength only should be specified. The latter group feel that very high bond strength (if attainable) might contribute to greater susceptibility of flex cracking.

Testing of heat seals. Testing of heat seals includes visual inspection for presence or absence of defects, physical tests for seal failure under peeling force, and for resistance to seal failure under hydraulic pressure applied in a pouch-burst tester. The latter test will detect abnormally weak laminates if seals are stronger than the pouch wall.

The pouch-burst test requires one end of the pouch to be open. This end is clamped over a water-injecting device. A thick transparent glass cover is then clamped shut at a fixed separation distance from the bottom of the tester. This prevents the pouch from swelling beyond the set distance. Water is introduced at a controlled temperature, and water pressure is increased at a specified rate until failure of one of the three seals or of a body panel.

Alternatively, the water is brought to a predetermined pressure and the pouch held at that pressure for a specified time, e.g. 60 seconds. With this test, surviving seals are examined for signs of creep.

Pouch-burst tests can be made before filling, after filling and closing, and after filling, closing and retorting. In the latter two instances the contents have to be removed. This is usually done by cutting off the bottom seal so that side seals and the very critical top seal can be tested.

Visual inspection of seals is required to identify potential sources of recontamination of the contents after processing by way of gross gaps, wrinkles, fold-overs, blisters, inclusions (foreign matter) and food particles. Examination of the final seal is of critical importance, as this is where the greatest chance for seal contamination exists.

Considerable effort is being spent in trying to develop devices for automatic scanning of seals to identify defectives. Most have depended upon detecting of differences in infra-red or laser beam reflection. In the absence of an effective automatic defect detector, the retort pouch lines require careful 100% visual inspection of seals.

Product fill and residual air testing. Control of fill is important in avoiding seal contamination and in avoiding excessive pouch thickness during retorting. Control of residual air is important for two reasons. Oxygen adversely affects the quality of some foods, and excessive internal air pressure will interfere with heat transfer as well as contribute to potential pouch swelling or bursting. The more overriding air needed in the retort to counter balance pouch internal pressure; the less efficient the heat transfer.

Residual air is measured by cutting open the pouch under water and catching expelled air bubbles in an inverted graduated cylinder (Steffe *et al.*, 1980).

Abuse and shipping tests.   Abuse and shipping tests are designed to simulate shocks and vibrations that occur in the distribution cycle.

Economics

The economics of the retort pouch depend a great deal on the frame of reference. Steffe *et al.* (1980) reported substantial energy savings for a pouch line. On a comparison of materials costs alone, the pouch plus printed overcarton plus shipper compares favourably to the cost of the can plus label plus shipper. The can once had an advantage, but as metals costs rise and plastics costs fall, the scales will tip increasingly in favour of the pouch.

On a comparison of total costs, including energy, warehousing, and shipping, the pouch looks even more favourable. Processing time is shorter, solids fill is greater per unit, empty warehousing is 85% smaller, filled packages about 10% less, and weight of the empty package is substantially smaller.

On a comparison of quality and convenience, the retort pouch is superior to the can. It is easier to open and the food can be reheated in the pouch. It is easier to dispose of, and the nutrient content of the food is higher.

On a comparison of capital investment for equipment, the pouch comes off second best where canning lines are already in place. For new installations, it might have a more favourable comparison, despite low speeds, where short runs are the rule.

On a direct labour basis the can is well ahead. Until pouch lines can be brought to a higher level of automation, particularly in the making and inspection of top seals, this will continue to be the case.

The market for the retort pouch

With the exception of Japan, the retort pouch has not enjoyed a pronounced success in the civilian market place. In the author's opinion this has been due to a lack of perceived value on the part of the consumer.

The US consumer has access to fresh, frozen, and canned foods and has a basic understanding of comparative quality, convenience, and price. It will take substantial consumer education to convince the shopper that retort-pouched food has sufficient advantages and values to justify its price.

The military, on the other hand, have a perceived need and are willing to pay what is needed to sustain the source of supply. It is to be hoped that the experiences gained in supplying the military market will carry the load, until further reductions in cost will enable penetration of the civilian market to a level sufficient to encourage large-scale commitment of capital.

Countries that do not have substantial investment in can manufacturing or in frozen-food processing plants, but rely on imported cans or frozen products, are more likely candidates for retort-pouch operations. Shipping costs of prefabricated pouches will be substantially lower than for cans and labour costs are lower, Manual or semi-automatic operations will be acceptable, quality requirements may be less stringent, and refrigeration in the home may be far less prevalent.

It has been estimated that the potential world market for the retort pouch will approach 100 billion units by the 1990s. Such a level will not be met unless there is a breakthrough in the civilian markets of developed countries or adoption by developing nations, possibly with government subsidies, to reduce the huge losses of perishable foods that occur after harvesting.

The evolution of plastic retortable packages

The foil laminated tray

Formed trays made from thick-gauge aluminium foil have been used extensively for frozen foods. The application of a heat-sealed lid to the smooth flange of such a tray seemed a logical step to a retortable tray. Retort-pouch stock was the most likely candidate. Presence of forming oils on the trays and difficulty in achieving seals led first to the coating of the tray and ultimately to the use of a heavier version of the retort-pouch laminate. In one version offered by Alusuisse, the aluminium is laminated to cast polypropylene on the interior side for heat sealing and to oriented polypropylene on the exterior side. Such a laminate can be forced into a female cavity mould by compressed air, vacuum, or a male die. The plastic and foil cold-draw without fracturing of the foil and without subsequent delamination. Multiple cavity-forming from a single strip is feasible. Product is readily filled from above, and the top web heat-sealed in place before separating the individual trays. Vacuumizing before sealing with optional gas-flushing is recommended, as is retorting in pressure-balanced retorts using water or steam cooks with suitable racking. Original web stocks required cutting open with a knife or scissors. In the last two years, peelable lid stocks have been developed which will survive retorting. (See discussion of retort pouch adhesive above.)

Retortable plastic materials

The original low-cost retortable polymers — polypropylene and high density polyethylene — do not provide sufficient barrier to oxygen to provide long-term shelf life at ambient temperatures of storage. Better barrier materials (polyester, polyacrylonitrile) are more expensive. The high-barrier materials (polyvinylidene chloride and ethylene vinyl alcohol polymers) are prohibitively expensive, and the exotic best-barrier materials, the fluoride polymers, are beyond consideration. New, better barrier polymers are beginning to come on the scene and will be evaluated (Table 1.4).

A technology now available offers promise. Multiple co-extrusion (Figure 1.2) permits the burying of a thin layer of more expensive barrier polymer

Table 1.4 Barrier properties of selected thermoplastics

Thermoplastics	Oxygen transmission cc.mil/100 in^2.24h.atm @ 73°F (23°C) and 75% RH, ASTM D-1434		Water vapour transmission rate g.mil/100 in^2-24h @ 100°F (38°C) and 90% RH, ASTM D-96
*Saran Resin F-278	0.02		0.02
	0% RH	100% RH	104°F (40°C)
**Eval F	0.01	1.15	3.8
Eval E	0.09	0.65	1.4
Nitrile Barrier Resin	0.8		4–5
Nylon 66	2.0		6–10
Nylon 6, biaxially oriented	1.2		10
Nylon 6	2.6		22
Polypropylene	150		0.25–0.7
Polyethylene Terephthalate (PET)	4.8–9		1.8–3.0
Rigid Polyvinyl Chloride	5–20		0.9–5.1
High Density Polyethylene (HDPE)	150		0.3–0.4
Medium Density Polyethylene (MDPE)	250		0.7
Low Density Polyethylene (LDPE)	420		1.0–1.5
Polystyrene	350		7–11

** Trademark of Kuraray Co. Ltd. for EVOH (ethylene vinyl alcohol) polymers.
 * Trademark of Dow Chemical Co. for polyvinylidene chloride copolymers.

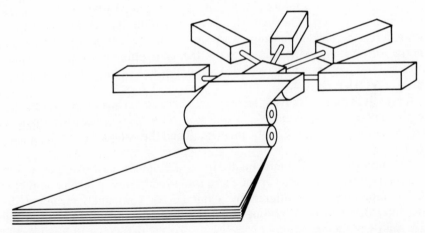

Figure 1.2 A multiple feed block coextrusion. Redrawn from brochure published by Ball Plastics Division, Evansville, IN 47712.

between two or more thicker layers of less expensive polymer. One of the best processes is the feed-block co-extrusion of multilayer sheet patented by Dow and licensed to other companies. With this process the number of polymer layers can be quite high. Reliable sources at Dow claim to have produced experimentally as many as 100 plies. An additional key to the success of this process is the use of molten adhesive 'tie layers' to cause the dissimilar polymers to adhere together.

Multi-layer barrier coextrusion melts can be blow-moulded, injection blow-moulded, injection-moulded, or cast into sheets. Sheets can be thermoformed or cut into billets and compression-moulded.

The competing barrier polymers, PVDC and EVOH, have advantages and disadvantages. EVOH can be tolerated in low percentages in scrap regrind. Utilization of regrind containing PVDC is not yet commercial. EVOH is an excellent oxygen barrier when dry, but the barrier is drastically lowered if the polymer is plasticized by moisture. PVDC proponents claim moisture will be driven into the EVOH during retorting. EVOH proponents do not regard this as serious, but American Can Co.'s 'Omni' Can (see later) is made from a co-extrusion that contains a desiccant in the tie layers to capture any errant moisture before it can affect the EVOH layer.

Retortable thermoformed plastic containers

The RTF (rotary thermoformed) process developed by Hercules co-extrudes multiple-layer sheeting directly on to a rotary drum where it is thermoformed into multicavities whilst the sheet is still above the melt temperature of the polypropylene. Cooling occurs during thermoforming, and the formed sheet can pass directly to a filling and lidding station prior to being severed into individual containers. The higher temperature results in essentially stress-free containers, thereby minimizing warpage during retorting. Hercules claim thermoforming at temperatures below the melt point not only introduces stresses (and tendency to warp) but accentuates loss of barrier characteristics of the EVOH layer.

A typical Hercules RTF container would have seven layers: polypropylene/regrind/adhesive tie-layer/EVOH/adhesive tie-layer/regrind/polypropylene. The polypropylene layers can be pigmented and the outside layer talc-filled, if so desired.

Thermoformed containers made from previously cast composite (co-extruded) barrier sheetings by the more traditional forming technique (below melt temperature) are offered by a number of companies including, for example, Ball Plastics Division, Metal Box Co. ('Lamipac'), American Can Co., and Continental Can Co.

A thermoformed rigid tub made by PLM of Denmark and called 'Ultra-lock' is made from polypropylene/PVDC/polypropylene, with an easy-open pull-ring lid sealed to the bottom by ultrasonic welding.

Retortable pressure-formed plastic containers

Rampart Packaging, a subsidiary of Shell Chemical, have developed a 'scrapless' method of container manufacture. Coextruded barrier sheeting is die-cut into billets. Scrap from this operation is reground and incorporated into inner layers of the laminate.

The billets are placed in a mould, preheated, and compression moulded into the container shape. Scrap is said to be about 9–10%, compared to 40% from traditional thermoforming processes.

Container lidding stock

Peelable heat-seal lidding stock for metal or plastic containers is available from several suppliers, and is typically fabricated by lamination. It may be of two types: (i) polyester film/adhesive/aluminium foil/adhesive/polypropylene film, or (ii) oriented polypropylene film/adhesive/aluminium foil/adhesive/ cast polypropylene film.

For the Campbell 'Cookbook Classics Soup Bowl', Reynolds Aluminium supplied a peelable fusion-sealed lid made from a lamination with heavier-gauge foil of sufficient thickness to be strong and puncture-resistant. Lids are pressure-formed and die-cut to recess into the bowl to reduce head space.

Retortable injection blow-moulded plastic containers

American Can Co. coupled a multi-layer co-extruder with an injection blow-moulding machine to produce a cylindrical seamless 'can' with sturdy flanges that will accept a seamed-on metal can lid. The lid can incorporates a ring-tab pull for easy opening. This can was introduced by Hormel in mid-1985 as the 'OMNI' (tm) Can. Because the process is scrapless, there is no need for regrind layers. Deeper draws can be achieved than by thermoforming processes. The container base is shaped to nest the lid of a can beneath, which permits stable stacking.

Because of the high temperature in forming above melt temperature, PVDC cannot be used for barrier. EVOH is suitable, provided desiccants are added to the tie layer adhesives to scavenge moisture, as previously described.

This container is claimed to offer substantial economies over metal cans, as well as attractiveness and easy opening. It will have somewhat slower heat transfer for properties during retorting, and retains the disadvantage of the cylindrical shape versus the thinner retort pouch.

Other candidates

The 'Let Pak', developed by the Akerlund and Rausing Division of the Swedish Match Co., appears to have shortcomings that may inhibit its acceptance. This is a three-piece 'can' made from a polypropylene/aluminium foil/polypropylene body with laminated 'plastic' ends welded to the body by a high-frequency electric process. The package has a tall rectangular shape, which contributes better 'cube' efficiency, but early examples tended to split open when dropped.

'Multi-therm' is a system developed by Alfa Star. It comprises a hydrostatic processor in line with a thermoform–fill–seal machine. Pouches or lidded trays are formed, filled, and sealed in continuous web. The web is fed through the processer, cooler, and dryer and then severed into unit packages.

Packages are barrier plastic (no metal), as microwave energy is used to heat the food. The web passes in sequence through the following:

(i)   A preheating stage in water at 80°C (176°F)
(ii) A pressurized heating chamber in water at 127°C (261°F)
(iii) A precooling stage in water at 90°C (194°F)
(iv) A cooling stage (water)
(v)   An air dryer.

Microwave energy is used in the pressurized heating chamber to heat the food. Because the packages are immersed in water, the microwaves cannot 'see' the transparent flanges. This is claimed to prevent overheating of seal areas. The presence of the water is also claimed both to speed up processing and to moderate temperature differentials within the food. Substantially shorter processing times, energy savings, and improved food quality are claimed by Alfa Star.

Still another concept, called 'Achilles', is offered for development by Alfa Star. This is limited to liquids and again uses a continuous web of plastic. The plastic is formed into a tube with flat (thin) cross-section. As this tube passes through the heating zone (heating method not specified), the liquid is fed through the tube at 10 times the speed of the tube. Because of the thin cross-section, sterilization is accomplished very quickly. After cooling, the web passes through a pouch-forming station, where liquid is allowed to accumulate and slow down to the tube speed. Cross-web seals are made through the liquid, and pouches severed. Alfa Star claim extremely short processing times at 155°C (310°F) temperature. Considerable more development effort is needed to perfect the Achilles and 'Multi-therm' processes which have so far been tested only in the laboratory. Alfa-Star are offering joint development effort licenses to potential customers.

Summary

Plastic containers with heat sealed lids will have the same problems as the retort pouch — proof of satisfactory seals and avoidance of splash or contamination of the seal areas. The method of filling and applying the lid with the container in horizontal position should reduce this hazard substantially.

Assuming that co-extruded barrier plastics and/or new generation resins provide adequate shelf life, elimination of metal foil from the laminates should reduce flex cracking, which is currently a matter of concern. This will permit use of microwave energy for heat-processing, as is now being explored by Alfa-Star.

The future of plastic retortable containers (and the retort pouch) for foods depends greatly upon the economics and improving technology of competing aseptic packaging. In the opinion of the author there will be room for both technologies in food packaging, but pharmaceuticals will remain with aseptic processes.

The greatest need for such packages will be in countries that do not have existing can-making or glass-jar making industries. Importing of plastic resins or co-extruded plastic laminates will be less costly than importing empty containers. Container making, filling, and closing equipment will require substantially lower capital investments.

Plastic containers will offer further savings in the distribution system, and shelf-stable foods will not require costly refrigeration.

References

Anon. (1982) Test Cycles for Small Size Flexible Retortable Pouches (Guideline to Container Manufacturers and/or Processors issued by Canned Products Branch, Processed Products Inspection Div., USDA/FSIS/MPITS, Washington DC 20250).
Anon. (1985) Retort technology pressures plastic can development. *Pkgg. Dig.* January (1) 165–170.
Anon. (1985) Applications for Alfa Star's multitherm process. *Pkgg. Strategies*, April 15.
Anon. (1986) New Wave in Soup Packaging. *Fd. Processing* **47** (1) 21–22.
Baccaro, L.E., Mihalich, J.M. and Hirt, R.P. (1985) Materials and composites for high-temperature barrier packaging. *TAPPI J.* **68** (1) 59–63.
Bannar, R., (1979) What's next for the retort pouch. *Fd. Engg.* **51** (4) 69.
Bertrand, K. (1986) The 6 hottest trends in packaging. *Packaging* **31** (1) 24–31.
Dilberakis, S. (1986) Barrier plastics: the next generation — container options snowball under domino effect. *Ed. and Drug Pkgg.* **50** (1) 8, 36, 37, 39 (see also 3, 4).
Foster, R.H. (1984) Comparative economics of composite barrier containers. Presented at 2nd International Ryder Conference on Packaging Innovations, Dec 3–5 1984.
Hannigan, K. (1984) Thermoformed from molten plastic: retort cup. *Fd. Engg.* **56** (8).
Labell, F. and Rice, J. (1985) The retortables — plastics that can 'Take the Heat'. *Fd. Processing* **46** (3).
Lampi, R.A. *et al.* (1976) Performance and integrity of retort pouch seals. *Fd. Tech.* **30** (2) 38–48.
LeMaire, W.H. (1985) This processing/packaging system uses...microwaves. *Fd. Engg.* **57** (6) 50–51.
Morris, C.E. (1984) Coming: retortable plastic cans. *Fd. Engg.* **56** (4) 58–59.
Morris, C.E. (1986) Plastics food packaging in review at Foodplas. *Fd. Engg.* **58** (4) 28–29.
Morris, C.E. (1986) PET material debuts in...spaghetti jars. *Fd. Engg.* **58** (4) 30.
Morris, C.E. (1986) The other challenges facing plastics...lidding. *Fd. Engg.* **58** (4) 32–33.
Morris, C.E. (1986) Commercializing food in barrier plastics — microwave cuisine. *Fd. Engg.* **58** (4) 37.
Morris, C.E. (1986) Plastic cup forms festive...table mold. *Fd. Engg.* **58** (4) 38.
Morris, C.E. (1986) What to do to shift from glass to barrier plastic. *Fd. Engg.* **58** (4) 40–41.
Rice, J. (1986) Contract packaging in retortable plastic trays. *Fd. Processing* **47** (4) 160–161.
Steffe, J.F. (1980) Energy requirements and costs of retort pouch *vs.* can packaging systems. *Fd. Tech.* **34** (9) 39–43, 75.
Swientek, R.J. (1986) Continuous sterilizing system handles plastic containers with heat-sealed closures. *Fd. Processing* **47** (1) 76–78.

2 Aseptic packaging

FRANK A. PAINE

Introduction

All techniques of food preservation are designed to prevent spoilage and limit changes, and thus impart 'shelf life' to the food. The most important methods of food preservation used by the food industry at the present time are:

(i)   Heat processing
(ii) Refrigeration
(iii) Reduction of available water
(iv) Preservation by chemical means
(v)   Control of the atmosphere surrounding the food
(vi) Curing and smoking
(vii) Fermentation.

Other more recent, and at the present time less important, methods are irradiation with gamma-rays or electron beam; the use of UV light; ultrasonics; and antibiotics. Of these methods, heat processing is by far the most efficient in terms of increasing shelf life, and aseptic packaging is probably the most likely to be developed in the next decade.

Heat processing

The details of the heat treatment used in any instance will depend not only on the nature of the food but also on the kinds of micro-organisms involved and whether other preservation methods will also be used. The first successful long-term preservation of foods was achieved by Appert's canning process in the early 19th century. Canning and bottling processes have one great advantage over all other methods, particularly when the working conditions are not favourable or are unhygienic, in that the food is not processed until it is inside its sealed container, and having been sterilized by heat and rendered free of all pathogenic organisms is protected from reinfection by the integrity of the seal on the can or jar.

In canning and bottling processes temperatures of 110–140°C are used to destroy or render inactive any micro-organisms that could cause spoilage and/or toxicity. In fact, total sterility is unnecessary and the complete destruction of all micro-organisms is seldom obtained. The cells and spores of micro-organisms differ greatly in their resistance to temperature, and a major factor in this is the acidity or alkalinity (pH) of the environment. Both cells and spores are generally most resistant to destruction by heat when the substrate is at or near neutrality (pH 6–8). Any increase in acidity or alkalinity hastens killing by heat, but movement towards more acid conditions is more effective than becoming more alkaline.

The processing time required therefore decreases as the pH moves from the neutral region in either direction, but more rapidly as it becomes acidic. Most of the heat-resistant pathogenic organisms (e.g. *Clostridium botulinum*) will survive longer at a pH above 4.5 than below this value. Foods with pH 4.5 or below are termed 'acid foods' and those above as 'less acid foods'. The processing of acid foods can be carried out at the temperature of boiling water and atmospheric steam, 100°C, in a relatively short time, but the less acid foods must be raised to higher temperatures, up to 130°C, for longer periods to eliminate the risk of botulism. In every instance the sterilization process is followed by rapid cooling.

Pasteurization

This is the term applied to the milder heat treatment given to some products to kill most, but not all, of the organisms present. Normally non-sporing pathogens are killed and the shelf life of the food is thus prolonged. Temperatures below 100°C, generally 60–75°C, are employed and the heating is by means of steam, hot water, dry heat or electric current. The product is rapidly cooled immediately after heat treatment. Pasteurization is used when:

(i)   A more severe treatment would harm the quality of the food
(ii) Pathogens require inactivation, e.g. in milk
(iii) The main spoilage organisms are not very heat-resistant, e.g. yeasts in fruit juices
(iv) Any organisms which survive the treatment can be controlled by other preserving methods, e.g. by chilling
(v)   Competing organisms need to be removed to allow a desired fermentation from an added starter organism, e.g. in making cheese.

Several other preservative methods are used to supplement pasteurization — refrigeration has already been mentioned. Alternatively, the product may be packaged in a sealed container to prevent reinfection, or packaged in an evacuated container to maintain anaerobic conditions, and a chemical preservative or a high concentration of sugar may be added. The duration of the treatment and the temperature depends on both the method of heating used and the product.

B

New trends in retorting

Retorting has long been the traditional method of sterilizing foods in-package. Cans and glass jars were originally treated in batch autoclaves by steam or hot water. (Figure 2.1) The batch autoclave is still an important piece of equipment for many products but during the last 50 years or so considerable improvements in energy saving and control equipment have been made. Continuous retorts were developed either as tunnels using hot air to sterilize, or rotating autoclaves fitted with rotary valves or hydrostatic towers using columns of water as transit pressure valves for the cans, which are moved through the process on chain conveyors.

The traditional method of sterilization involves filling the food into its container which is then hermetically sealed before heat treatment in the retort. The continuous-flow system combined with aseptic filling was developed from the retort system because increasing the temperature in combination with a shorter holding time has the same lethal effect on micro-organisms while reducing the adverse effects of other possible chemical reactions on the food. Moreover, in a continuous-flow process the product temperature can be raised and lowered more rapidly than in a batch process. Hence the time distribution for the several parts of the system will be more even.

The time–temperature programme of any process is determined by the lethal effect required. As already mentioned, the possibility of other chemical reactions is decreased at higher sterilization temperatures and it is therefore advantageous to use as high a temperature as possible. For technical reasons, heating the product to the temperature required for sterility and then cooling

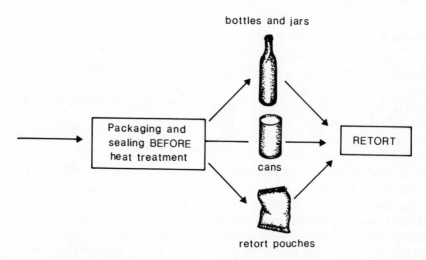

Figure 2.1 Principles of thermal sterilization: traditional in-can method.

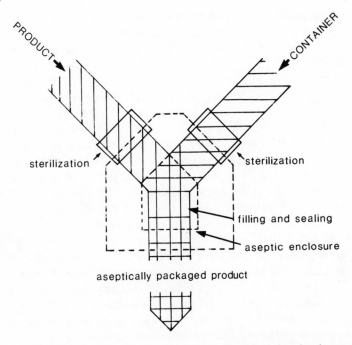

Figure 2.2 Principles of thermal sterilization: aseptic processing. Technique involves sterilizing product and package separately. They are then brought together in a sterile (aseptic) enclosure.

cannot normally be done instantaneously, and the times to heat up and cool down are also of significance in both the lethal effect and any chemical effects. These times must be kept as short as possible.

The basis of sterility calculations

All calculations of the time/temperature conditions required to achieve sterility in a thermal process are based on the same principle. The percentage of the bacteria killed in unit time as a proportion of the total present at the start of the operation remains constant, i.e. the process is similar to a first-order reaction. We are interested, however, in the number that survive, and this will decrease exponentially. Based on this precept we can derive a relationship between heating time, number of bacteria, and temperature which will allow the calculation of the conditions needed to achieve 'commercial sterility'.

For many products this calculation gives a value between 5 and 12 minutes at 121°C for the coldest point within the package. The dimensions, particularly the thickness of the package, determine how long it will take the coldest point to reach the sterilizing temperature. For typical cans and jars these times are relatively long (15 minutes to over an hour in some instances.)

The benefits implicit in a continuous sterilization process can only be fully realized if it is combined with aseptic filling and packaging. Aseptic packaging implies sterilization of the package or the packaging material and filling the cold, commercially sterile food into it under conditions which prevent re-infection during the filling, closing and sealing operations. The integrity of the closure or seal must also be maintained during transport and distribution.

Thus aseptic packaging technology opens up opportunities for better packaging materials and systems. Unlike conventional canning, the aseptic process, involving as it does a separate sterilization of product and container, causes less thermal damage to the product and less stress on the packaging. Besides improving product quality, it allows the use of other materials than the traditional metal can or glass jar. Packages of laminated paperboard or plastic containers of various shapes may be used instead of cans and jars, and this usually reduces the cost of the materials (Figure 2:3).

This is not to say that cans and jars are not used in the aseptic field. They provide good protection and their closures and seals are of high integrity. Moreover, their good thermal resistance makes it possible to sterilize them using steam or hot air, which is not possible with most paperboard or plastics packaging. However, cartons and plastic containers generally have a cost advantage and can also be produced in-house on relatively simple form–fill–seal packaging equipment. Their sterilization, however, requires the use of chemical agents such as hydrogen peroxide, which has given rise to questions of effectiveness and the possible presence of sterilizing agent residues after treatment. These considerations delayed the use of such methods for a long time in the USA until the FDA had ruled on the matter.

Let us now examine the four major requirements in some detail. They are, as we have already shown, sterilization of the food, sterilization of the package or packaging material, maintenance of sterile surroundings during

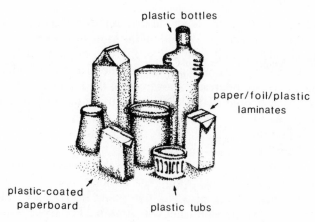

plastic bottles

paper/foil/plastic laminates

plastic-coated paperboard

plastic tubs

Figure 2.3 Aseptic techniques allow the use of other materials than the metal can and glass jar.

forming, filling and closing, and the production of seals (or their protection) of sufficient integrity to prevent reinfection during distribution.

Product sterilization

Producing a sterile product in a continuous process involves three steps: heating it to raise the temperature to the desired level; passing the product through a temperature-holding section for a predetermined time; and cooling the product as rapidly as possible to a temperature of 35°C or less prior to filling.

The type of heat exchanger used will be determined by the nature of the product, e.g. the viscosity of the liquid phase and the amount and size of any solid matter. In any continuous process the size of any particulates is limited to about 8–10mm diameter. Larger particles need special equipment, such as the APV Jupiter (see later). Ideally the temperature used would be high enough to obviate the need for a holding section, but this is never practical for several reasons. First, viscous products are difficult to heat uniformly; second, the food may contain heat-resistant enzymes which are more likely to survive the process; and third, the presence of particles makes it important to have a holding section to allow time for equilibration. Also, the shorter the holding section the more difficult the control becomes.

Table 2.1 Examples of sterilizing by continuous-flow systems

Method of heating	Company or process name	Observations
Indirect		
Tubular	Cherry-Burrell (USA)	
	Crepaco Inc. (USA)	
	Rossi & Catelli (Italy)	
Plate heat exchangers	Alfa-Laval Steritherm (UK)	Maximum viscosity
	APV. Co. Ultramatic (UK)	depends on product and
	Schmidt-Bretten (FRG)	temperature in cooler
Scraped surface exchangers	Alfa-Laval Contherm	Max. particle size 25mm
	Cherry-Burrell Votator	
	Crepaco	
	Fran Rica (USA)	
Direct		
Steam injection	Alfa-Laval VTIS	
	APV Uperizer	
	Cherry-Burrell	
	Crepaco	
	Rossi & Catelli Stematic	
	Steriglen Systems Pty. (Australia)	Max particles 35mm
Steam infusion	Crepaco Ultratherm	
	DASI Industries Inc. (USA)	
	APV Jupiter	Max particles 25mm

The equipment used can be classified according to the method of heat exchange used (see Table 2.1).

Indirect heat exchanges.　　The product and the heating medium are separated by a wall, the heating surface. The heat is supplied by steam or hot water. There are three types: tubular heat exchangers, plate heat exchangers, and scraped surface heat exchangers.

Direct heat exchange.　　The product and the heating medium are in direct contact. The only medium applicable to food is steam, which condenses in the product. There are two alternatives:

(i) Injection of steam into product — the food is the continuous phase in the mixing device and steam is injected into it
(ii) Infusion of product into steam — the steam is the continuous phase and the product is injected either as droplets or as a film.

Electrical methods.　　Dielectric heating can generate heat in the product and there is no heating surface. Such a method could be very attractive, especially for viscous foods. So far no commercial equipment is available, but development work is under way and some patents have been applied for.

Heating by mechanical friction.　　Here again the heat will be generated in the food itself. Energy is supplied as mechanical work and the source is electricity. A commercial machine is available from a French development company.

The first developments were designed for milk, and the calculation of heat transfer and pressure-drops involved little problem. When more viscous products are involved, calculation and design become more complicated, but if only one product is involved a good technical solution is almost always possible. Viscous products provide problems which result in increased pressure-drops and a decrease in heat transfer efficiency. Greater heating-surface area and longer residence time in every section of the equipment is necessary.

In recent years the aseptic packaging of products containing larger particles has become of great interest. Two problems arise. The first relates to the mechanical behaviour of the particles and the second to the time taken for the centre of the particles to reach the required sterilization temperature.

The problem of the mechanical behaviour of the particulates requires special designs for valves, pumps, heat exchanger pipes, and so on, to avoid clogging on the one hand and damage to the particles on the other.

The time–temperature development of the particle centres is a heat-transfer problem. In a liquid, heat is transferred mainly by convection, while in solids the transfer is by conduction, which is much slower. The heating, holding and cooling sections must be designed therefore to deal with the delay between the liquid reaching the temperature and the transfer of heat to the centres of the solid lumps. And, of course, the size and uniformity of those lumps will be of paramount importance.

Of the indirect heating methods available, only the scraped surface exchangers are capable of handling liquid products containing particles. Equipment designed originally for very viscous liquids has been modified to allow the passage of particles, and manufacturers claim that cubes of 20–30mm can be treated without significant damage.

The direct heat-exchange methods (direct mixing of steam and product) are probably of greater potential. Two systems are used, both of which handle the liquid and solid phases individually and sterilize them separately. In each instance the liquid is treated by one of the methods given above.

The APV Jupiter system treats batches of the particles in a rotating jacketed double-cone vessel. Steam is fed into the vessel and efficiently mixed with the particles by rotating it. The jacket is also heated by steam, and it is claimed that a temperature of 130°C in the particles can be achieved within two minutes. Cooling is carried out by passing cold water through the jacket. A number of commercial units are in operation.

The Steriglen system heats the particles with steam in a vibrating fluidized bed. The maximum size of particle is about 35mm. To give fast cooling, sterile nitrogen at −190°C is used. A commercial unit is operating in Australia.

One of the problems in developing equipment for handling particles lies in the design of pumps and valves which allow large particles to pass without damage, while still maintaining a satisfactory rate of operation.

To obtain the best quality in the finished product we need to reduce the heating and cooling times of the particles to a minimum. Heat transfer from the outside to the inside of a solid particle is effected mainly by conduction, so that if the particles of food approach the same size as the thickness of the package, the difference between consumer-size aseptic packages and retort pouches will become negligible.

Sterilization of the packaging

The principal requirements of any method of sterilization for the package or packaging material are that it should be continuous, rapid and economical, as well as safe and without any side-effects or environmental risks to operators. The systems currently in use may be classified into physical, chemical, and processes using a combination of these. *Physical processes* include sterilization by steam or hot air, and the heat or energy utilized in the process of making the material or container. *Chemical processes* utilize such chemicals as chlorine, peracetic acid and hydrogen peroxide. Hydrogen peroxide in conjuction with heat, and at much lower concentrations in conjunction with UV radiation are the main *combination techniques*. Processes using ionizing radiation or ethylene oxide have also been studied.

To reduce the risks of infection, the sterilization of the container or the packaging materials is not carried out until just before filling; indeed, in some systems the two processes have been integrated. The oldest method still in use is exposure to superheated steam or hot air which, because of the high temperature involved, limits the choice of packaging to glass containers or cans in the main. Treatment of plastics and most flexible packaging is therefore mainly carried out using chemicals or combination systems.

A better way to use heat sterilization is to take advantage of the heat generated during the manufacture of the packaging. The temperatures reached during melting and extruding plastics are usually well above that needed for sterilization. Therefore, immediately after forming, plastic containers, films and laminates are quite sterile. The Bottle-pack system, for example, fills the container immediately after forming while it is still in the mould. In another system (Siderac) the bottle is passed to the filling head under sterile conditions. A third system produces a plastics bottle which is sealed, and hence the interior is kept sterile until required to be filled. Such containers can be stored for long periods without becoming infected. When ready for filling they are handled in a second set of equipment which sterilizes the area around what will become the closure, before cutting it open under sterile conditions, filling and closing it.

A similar principle is adopted with co-extruded sheets in the Erca process, where a protective peelable layer is provided both on the material used for the thermoformed containers and the lidding material. Just prior to processing the protective layer is peeled off, the containers are formed by heat and filled, and meet the lidding film which has also just had its protective layer removed. Thus the contents only contact the inner layers which have remained sterile since the laminate was produced.

Apart from sterilization during manufacture, two other methods of solving the problem have been devised. The most important method for presterilization of packaging material in aseptic systems is by the use of chemicals, and hydrogen peroxide is the most widely used. Different techniques are used for preformed containers and reel-fed material which is then formed into the required shape.

Web sterilization

This is a continuous process and is mainly carried out using hydrogen peroxide and heat (IR, hot air or heat conduction). The packaging material passes into and through a bath of the chemical, and surplus liquid is removed by rollers. The container is then formed into a tube (for example) by heat-sealing longitudinally, and the remaining peroxide is removed by a radiant heater inside the tube (Tetra Pak and Tetra Brik). The same principle is employed in the Benhil–Formseal system where cups and lids are both formed from reels. Here the peroxide bath is followed by suction drying to remove excess.

Container sterilization

The material from which containers are to be made is sometimes sterilized by chemicals, e.g. ethylene oxide, before it leaves the convertor. The container is then formed at the packer's plant and sterilized with a spray of peroxide, followed by blowing with sterile hot air to remove excess. (Pure Pak and Combibloc are handled this way). A similar procedure at the packer's plant is also used for plastics tubs.

Some fundamental aspects: reducing the bacterial load

This subject has been studied at the Fraunhofer Institute for Food Technology and Packaging, Munich, by G. Cerny and his co-workers, and some consideration of the findings is worth making here.

It is well established that most of the microbial organisms found on packaging materials are held by weak forces. Indeed, most microbes are attached to dust particles by electrostatic charges. Hence they are removed relatively easily and this was investigated. It was found that micro-organisms, whether single cells or cells attached to dust particles, could be removed from plastics films by ultrasonic treatment in sterile water tanks. With preformed polystyrene tubs, blowing with sterile air also gave good results.

These treatments do not give complete sterility, but the reduction in the microbial load can be up to 10 000 times, which gives considerable security for the presterilization method to follow.

The organisms most resistant to treatment with hydrogen peroxide are bacterial spores, in particular *B. subtilis*, and this organism has been generally accepted as the test organism to use for peroxide technology. In all aseptic systems, sterility should be achieved rapidly, hence the chemical treatment is generally combined with heat. Reel-fed packaging materials are usually passed through the hydrogen peroxide at a temperature of up to 80°C, but sometimes at ambient temperature, when a small amount of wetting agent is added to the bath. The material is exposed to IR radiation or hot sterile air to vaporize and remove any excess peroxide. Even where wetting agents are used, preformed packages made from essentially hydrophobic materials such as polystyrene, or packages with polyethylene coatings on the inside, are covered only to about one-third of the surface area with small drops after spraying and the more reactive vapour phase produced by heating completes the sterilization.

A further way to reduce the initial bacterial load on the surface of packaging materials is to use UV radiation. The greater proportion of micro-organisms are inactivated within a few seconds if high-intensity lamps are employed (Figure 2.4), but after this the effect is slow, because the remaining spores are shielded by dust particles or the effect of cell aggregation.

Studies in the UK, Japan and Switzerland, confirmed in Germany, have

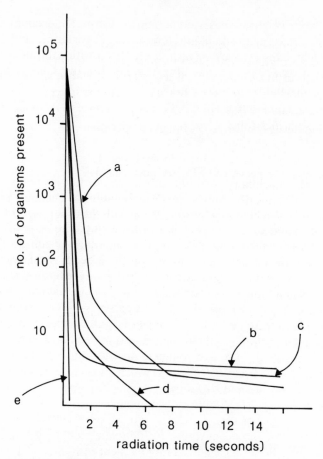

Figure 2.4 Inactivation of different micro-organisms by intense UV C-irradiation (UV intensity at 254nm = 30mW/cm^2). (*a*) *Aspergillus niger*; (*b*) *B. stearothermophilus*; (*c*) *B. subtilis*; (*d*) *P. frequentans*; (*e*) *S. cerevisiae* and *Rh. graminis*.

shown a synergistic effect using UV and hydrogen peroxide in combination. Moreover, the concentration of peroxide solution needed is considerably reduced.

Materials may also be sterilized by exposure to gamma-rays or electron beams. The technique is employed on laminate bags used in the 'bag-in-box' technique. Cerny *et al*. have examined this and found that a dosage in the range 15kGy (1.5Mrd) is needed, even higher for large-volume bags. At this level some plastics, such as LDPE, are affected, resulting in possible odour problems and heat-sealing and migration complications. More study is necessary before such techniques can be commercially recommended. We have already mentioned that the use of superheated steam limits the materials that can be used as packages, because plastics such as polyethylene are

softened and/or melt at these temperatures. Cerny has examined polypropylene, which has a much better temperature resistance, and found that
polypropylene cups can be successfully sterilized with saturated steam. It is
possible that other plastics may also be treated in the same way, provided
they can withstand the thermal stress.

Maintaining sterile conditions during plant operation

Presterilization of the packaging machinery

Generally, the machinery is sterilized by steam or hot air before processing
commences. Chemical methods, or a combination of chemical and thermal
processes, may also be used. In comparison with the other parts of the aseptic
process, the techniques of sterilization of plant and equipment has been little
studied, and could be regarded as the weakest part of the system. Essentially,
we must obtain and maintain sterility of all parts of the plant that make
contact with either the sterile product or the presterilized packaging before
these have been brought together and the product securely closed in its
container.

   Contamination with micro-organisms can only be excluded by designing the
plant efficiently, adopting sound and hygienic engineering principles, and
maintaining a rigorous regime of control to ensure that the plant is operated
properly. *Hygiene is paramount in all considerations of design.* The design
should take three main considerations into account:

(i)   It must be able to be effectively sterilized before start-up
(ii) It must completely isolate the product from the surrounding environ-
      ment and the operating personnel
(iii) It must permit product to flow without any risk that material could lodge
      during production so as to give rise to microbial growth.

As an example, a description of the operation of the Metal Box Freshfill
system has been described by Herbert. This uses a software package which,
combined with its own microprocessor, monitors plastic pot by pot all the
parameters which contribute to asepsis. To start up, the operator activates a
switch which initiates machine sterilization. Steam is supplied to the
UHT/filling machine interface, the product lines, surge tanks, fillers and
filling chamber. Thermometers are strategically placed in the system to
monitor temperature, and when the sterilization temperature (130°C) has
been maintained continuously for 20 minutes the steam supply is cut off. As
steam condenses, absolutely filtered air is admitted so that a positive pressure
is maintained in the product lines, fillers and surge tanks, and a continuous
flow of sterile air, or inert gas if desired, passes through the filling chamber
and the undercover-gassing chamber. The software programme ensures that
these positive pressures are maintained throughout the production period.

As soon as the lines and tanks have cooled, and sterile product is available, as signalled by the UHT controller, the product is admitted to the surge tanks. Level and pressure are controlled in these throughout the operation. The conveyor can now run, but before pots enter the machine the microprocessor confirms that all critical factors concerning sterility are correct. This checks:

(i)   The peroxide sprays
(ii) Temperatures of sterile hot air for drying pots
(iii) The pressure and flow rate of air in the lines
(iv) Temperature and functioning of the lidding foil section
(v)   Heat seal temperature
(vi) Correct operation of conveyor cleaning and sterilizing equipment.

The control then monitors the plant, invoking stopping and alarm mechanisms if and when the system moves out of control. When production ceases, a clean-in-place system isolated from the UHT plant operates automatically.

Kilsby and Metcalfe suggest that the requirements for aseptically packaged foods are different to those for 'in-package' sterilized food, such as canned goods, and the only rational attitude processors can adopt is to try to exclude infection completely. They advocate what they call 'challenge testing' of the plant by exposure to micro-organisms at the critical areas. The safety of the process can only be assured through 'on-line' preventive control, not by 'off-line' end product testing.

The production (and protection) of satisfactory seals and closures

Any aseptic system must be capable of closing and/or sealing the package hermetically to maintain sterility during handling and distribution. The integrity of the closure and seal is therefore of paramount importance. Unlike in-package sterilized foods, where the container is cooled after sterilization with water which is a major source of inflecton, aseptic packages are generally filled cold and require no cooling. The integrity of the heat-seals used in most aseptic systems is principally influenced by the efficiency of the sealing system used and by contamination of the heat-seal area with product.

Provided a satisfactory seal or closure is produced or applied, the secondary packaging (i.e. the cases into which the units are packed for transport) must prevent deterioration and damage during handling and storage.

Systems available and possible developments

The Dole system was the first developed and has, of course, been on the market longer than any other. In essence, it is an on-line sterilization system which uses hot air or supersaturated steam to sterilize cans which are filled in an area kept sterile with steam or hot air. The cans are closed on a doubleseaming unit, with lids sterilized in the same way (steam or hot air). Originally used for milk, it has more recently been used for packaging institutional products such as cheese sauce, puddings and desserts in individual servings, using two-piece drawn aluminium cans.

The real market success for aseptic packages began once flexible materials became available. The market can be divided into three according to the size of container: (i) packages, mainly flexibles up to about one-litre capacity for direct household consumption; (ii) bag-in-box packages up to five litres for institutional use; and (iii) drums up to 250 litres for the industrial market. All three areas are available for liquid products containing no solids. Food products containing particles have a more limited range. Table 2.2 gives some idea of the areas in which particular packaging materials are employed.

The first important aseptic systems were produced for milk, and the development of these from the original Tetra Pak and Pure Pak are described in some detail in Chapter 6. Both are extensively used for milk and fruit juices, as are the Combibloc, Elopak and other cartons (see pages 87–91).

Other systems using packages other than cartons have been developed for various products, principally using plastics, and Table 2.3 lists some of the most interesting. As the table indicates, they can be divided into five main types:

(i)   Preformed containers, principally tubs and cups, which are sterilized, filled and closed with a heatsealed lidding, usually of coated foil
(ii) Vertical pouch-forming systems similar to the Tetra Pak principle
(iii) Thermoform/fill/seal systems fed from the reel
(iv) Systems starting from plastic beads which form a container via extrusion, and blow a container from a parison which is subsequently filled and closed
(v)   Bag-in-box systems.

Each has its own advantages and disadvantages, and selection of the most suitable for any specific purpose depends on several factors, including cost

Table 2.2 Types of package and material used in aseptic systems

Package type	Packaging material					
	Steel	Aluminium	Glass	Plastic film	Plastic container	Laminates
Drums, tanks	x					
Cans	x	x				
Bottles					x	
Jars			x		x	
Cups, beakers and tubs					x	
Pouches, sachets and bags				x		x
Cartons						x

Table 2.3 Aseptic packaging systems

System	Company	Method	Packaging sterilization	Products
DOG Aseptic	Gasti	Preformed	H_2O_2 spray or steam	Dairy
FreshFill	Metal Box	Preformed	H_2O_2 spray	
Prepac	Prepac	Pouch	H_2O_2 dip	Milk
DRV 13 ST	Thimmonier	Pouch	H_2O_2 dip	
DMR 200ST	Bosch	Pouch	H_2O_2 spray	
Servac	Bosch	Thermoform, etc	H_2O_2 spray	Puddings, cream
Conoffast	Continental, Erca	Thermoform, etc	During making	
Asepack	Benco	Thermoform, etc	H_2O_2 dip	Yoghurt, desserts, cheese
Bottle-Pack	Rommel	Blowmoulding	During making	Pharmaceuticals, milk
Serac	Serac	Blowmoulding	During making +H_2O_2 spray	Milk, cream
Siderac	Siedel	Blowmoulding	During making, +H_2O_2 spray or UV	Milk, juice, water
Totalpac	C.P. Remy	Blowmoulding	During making +H_2O_2 spray	Milk, water
	Scholle	Bag-in-box	Gamma-radiation	Acid products, banana purée
	Franrica	Bag-in-box	Gamma-radiation	Tomato products
	Liqui-box	Bag-in-box	Gamma-radiation	Acid products

and convenience. To be successful, the product characteristics and the shelf life required must be integrated with the packaging features to meet customer needs.

One advantage of the aseptic system is that the product can be sterilized and shipped to a remote filling and packaging point in bulk. The product can be kept in the bulk container until the unit packages are required, and there are some bulk systems which can handle viscous and particulate products which may be regarded as difficult. The packing station here does not need to have heat-processing equipment, but only apparatus for transferring the aseptic product from the bulk container and filling it under aseptic conditions.

The potential advantages of presterilization followed by aseptic packaging over in-container sterilization may be summarized as follows:

(i) There is a reduction in the heat denaturation during the sterilizing process
(ii) Quality of product is always better and no longer dependent on the size of the package
(iii) Much greater range of shape and type, and lower-cost packages

(iv) Package size ranges from small pots through drums, bags-in box, to tankers

(v) Much reduced energy cost.

The desire of almost all food-processing companies is to be able to use high-quality, low-cost raw materials from all over the world, store them from different times of harvesting and then process them with minimum loss of quality, low energy costs and easy packaging providing convenience and integrity to their customers. Aseptic packaging permits this.

References and further reading

Atherton, D. *Aseptic Processing*. Study to establish the capabilities and limitations of available machinery for aseptic processing and packaging of foodstuffs. Campden Food Preservation Research Association (CFPRA) Technical Memorandum 270, May 1981.

Burton, H. Aseptic packaging. *Soc. Dairy Tech. Symp. on UHT Processing of Dairy Products*, May 1969.

Hahn, G. Co-ordinating the sterilization of convection heated food products. *Food*, November 1982, 34.

Hallstrom, B. Process developments and their impact on packaging. *Euro Food Pack, Int. Conf. on Food Packaging*, Vienna, September 1982, 231.

Rose, D. *Guidelines for the Processing and Aseptic Packaging of Low-Acid Foods*. Technical Manual No. 11, CFPRA, 1986.

Symposium, *Aseptic Packaging*, April 1983, CFPRA Proc., Stratford upon Avon.

Symposium, *Aseptic Processing and Packaging of Foods*, The Swedish Food Institute (SIK), University of Lund, September 1985, especially the following papers: A.C. Hersom, Technical and scientific arguments (9); A. Astrom, Aseptic packaging (139); D.A. Herbert, Aseptic packaging in plastic materials (159); G. Czerny, Basic aspects of sterilizing packaging materials (166); B. von Bockelmann, General aspects of aseptic packaging in flexible containers; I.J. Pflug and L.G. Zechman, Microbial death kinetics in the heat processing of food: determining an F-value (211).

3 Modified atmosphere packaging

RICHARD INNS

Introduction and history

The use of controlled atmospheres to preserve food is well established. In the 1930s chilled fresh beef, stored under carbon dioxide, was being shipped from Australia and New Zealand to the UK. However, until comparatively recently such methods were confined to bulk supplies of meat and fruit. In the last decade, fresh food sales in supermarkets have increased considerably, resulting in changes in both distribution methods and the associated storage conditions. Most fresh and many chilled foods are now prepared and packaged in central depots from where they are distributed to the retail outlets. Food must therefore remain 'fresh' throughout a longer distribution chain and still allow a reasonable 'shelf life' in the retailer's premises. An increasingly competitive market also demands that supplies are available throughout the week, including Monday morning and Saturday afternoon — periods in the past when shelves became depleted because of possible deterioration over the Sunday closed period. Modified atmosphere packaging offered the possibilities of achieving a significant increase in shelf life without losing the description of 'fresh food.'

Thus, simply stated, Modified Atmosphere Packaging — MAP — is a process by which the shelf life of a fresh product is increased significantly by enclosing it in an atmosphere which slows down the degradative processes such as the growth of microbial organisms, whilst enhancing some beneficial actions such as retaining the desirable red colour of meat. The term 'controlled atmosphere packaging' is also sometimes used, but is an inaccurate description; the atmosphere inside any permeable package will change with time as gases diffuse into and out of the package at different rates, as well as gases being absorbed and given off by the food in many instances. The description should be reserved for truly controlled storage, such as the gas storage of apples where the conditions are maintained at specific levels over the period concerned.

As already mentioned, the concept is not new, and claims for the original invention go back many years. Flushing with nitrogen has long been employed in preserving fatty foods; vacuum packaging is a form of MAP, and the use of sterile air in aseptic packaging machines might also be so described. The first significant packaging trials, with individual portion packs rather than bulk supplies, were made in the late 1950s when vacuum packaging for meat, fish and coffee were first introduced. Experimentation then extended to gas flushing with nitrogen in the early 1960s. Much of this development took place in West Germany and Denmark, a marked impetus being provided by the research and promotional efforts of the larger film producers such as Hoechst and machinery manufacturers such as Multivac. This early research showed that for the best results temperature control to around 2°C was essential for many products.

Thoughout the 1960s and 1970s, some unspectacular progress was made in Europe, the greatest applications being the vacuum packaging of meats and cheese and the gas flushing of ground coffee. In 1981 a breakthough occurred. One of the most prestigious of retailers, Marks & Spencer, introduced MAP for its complete range of fresh meats. These were packed in gas-flushed transparent plastic trays made on thermoform-fill and seal machines. This was rapidly followed by MAP for fresh fish, cooked meats and bacon. The market has since been growing at better than 20% per annum.

Few new technologies announced with a fanfare of trumpets are truly new, because the technical concept is at first not economic. Eventually the supporting technologies move along so that the original concept becomes commercially, as well as technically, feasible, and then it takes off. This has happened with modified atmosphere packaging. The development in barrier technology in particular was the essential step forward.

Reasons for the use of modified atmosphere packaging

Why use MAP? First of all, it is about quality and the need to recognize that in preserving fresh foods we are dealing with products that have been parts of living organisms and retain at least some of the properties of living tissue for their entire existence in the 'fresh' state. Equally, chemical changes in foods and spoilage by microbes are activities whose rates can be controlled or stopped entirely by preservation processes. Heat sterilization, as used in canning, is one example of a preservation technique which destroys all significant microbial action, whilst freezing slows down chemical change and microbial activity and is capable of stopping them virtually completely if the temperature is reduced to below −40°C.

From the consumer's point of view, however, these conventional preservation techniques have some disadvantages. Straightforward *chilling* produces only very marginal effects on the shelf life. *Freezing*, even under the best conditions, will induce some changes in the product, particularly when thawing takes place. In any event, frozen foods have come to represent, in the minds of many consumers, the opposite to fresh; while *heat treatments* produce inevitable cooking changes which are not always regarded as beneficial by the consumer.

Modified atmosphere packaging seeks to maintain the elusive and ill-defined condition we associate with 'fresh' produce. We all know what we mean by 'fresh food', at least in the sense of a personal definition, but we would probably be hard-pressed to produce anything like an agreed workable definition. No definition will be attempted here, except to state that we are talking about a value judgement, based on taste and texture after preparation for the table and which often includes cooking. Thus the requirement for some form of processing does not invalidate the description. The majority of consumers would probably regard pasteurized milk as 'fresh', as they most certainly would both chilled meat and fish; but freezing both the last two below −18°C and selling them from a freezer cabinet would result in the loss of the 'fresh' connotation.

Thus modified atmosphere packaging fits into the important area of pre-servation where shelf life is extended without the loss of those important and elusive properties which constitute freshness in the consumer's mind, and therefore move the product into a premium bracket. As premium products mainly have short shelf lives, processes which extend this even by a few days are of great value to retailers by virtue of the reduced wastage, reduced costs for shelf filling, and in certain instances, the ability to offer products that might otherwise be obtainable only from a specialist outlet.

Technical requirements

The range of products is seemingly almost limitless, and extends well beyond the varieties of meat, fish, cheese and pasta products that we have been accustomed to seeing. Some fairly random examples of publications include a German study on French fried potatoes, work from Japan on peanut butter, American work on bulbs to both preserve and modify their flowering habits. In the USA, yeast has been so packed, and in Scotland, smoked salmon. Before we examine in some detail the types of atmosphere used for different products let us consider the action of the gases alone.

Three principal gases are used: carbon dioxide, nitrogen and oxygen. There is a possibility that carbon monoxide may become significant in gas mixtures, but for the moment it has limited use.

Carbon dioxide, which has a powerful inhibitory effect on bacterial and mould growth when present in concentrations above about 20%, is the most important gas. The manner in which it inhibits growth is not fully understood, but King and Nagel (1975) showed that some anaerobic and faculative species are also inhibited, indicating that the effect is due to more than just the exclusion of oxygen. High concentrations of the gas may lead to discoloration and a sharp acid taste in some foods, because carbonic acid is produced by dissolution of CO_2 in the water contained in the food. This may also lead to in-pack drip. A further problem consequent upon the solubility of the gas in both fatty and aqueous phases in the food is that of pack collapse, which is countered by adding a suitable balance gas.

Nitrogen is inert, tasteless and is virtually insoluble in water. By excluding oxygen it can inhibit the oxidation of fats (lipids) and reduce the possibility of mould growth.

The presence of *oxygen* is generally to be avoided with many products because its presence causes oxidation and enhances the growth of aerobic bacteria. However, it is a necessary means of sustaining the basic metabolism of respiring fruits and vegetables and is instrumental in maintaining the red colour of fresh beef.

In fresh foods, the most important deterioration factors are *oxidation* and

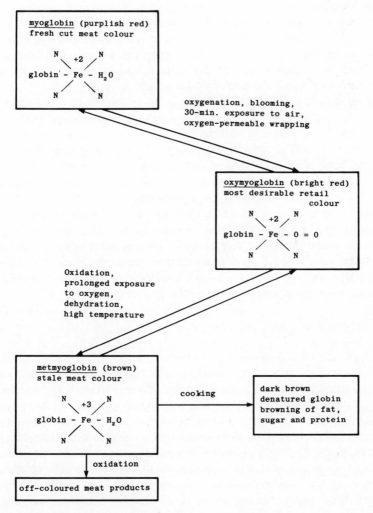

Figure 3.1 Colour changes in fresh meat. After Niven, C.F. (1951), Influence of microbes upon the colour of meat, *Amer. Meat Inst. Found. Bull.* **13**.

the *growth of microbial organisms*. Lipid oxidation, giving rise to rancid odours and flavours, can occur even when as little as 0.1% of fatty acids are present in the food together with oxygen. If the oxygen level in a pack can be reduced below 2% then the production of rancid odours is greatly retarded. Normally this is done by displacing the oxygen by nitrogen rather than carbon dioxide, which would be dissolved in the fatty tissues. The nitrogen-flushing technique for this purpose was used long before the 'modified atmosphere packaging' term was invented.

The situation becomes more complicated where red meat is concerned. Here it is desirable to increase the level of oxygen, because of the complex reaction mechanisms of the myoglobin pigments present. The transition from red to brown in meat is caused by the change of the bright cherry red oxymyoglobin colour, which appeals to the purchaser, into the brownish metmyoglobin (Figure 3.1). When the oxygen content of the atmosphere is raised to between 60–80%, this desirable colour is maintained for much longer, thus increasing the period of saleability.

To summarize, *oxygen* is used primarily to sustain basic metabolism and prevent anaerobic growths, *nitrogen* is used as a chemically inert gas to prevent oxidation, rancidity and pack collapse, and *carbon dioxide* inhibits bacterial and mould growth. Equally, as we have seen already, each of the gases can have negative effects in certain circumstances which need to be balanced against their positive actions; in particular there are optimum levels at which each needs to be used for a particular food.

Further elaborations occur when gases are first used within a package or process to produce desirable preservation effects and are subsequently flushed out with an inert gas. For example, carbon monoxide, ethylene oxide and sulphur dioxide gases have been used to reduce or block the actions of enzymes in some foods, and are then flushed out with either nitrogen or carbon dioxide to provide long-term storage.

Safety

Naturally, in the field of toxic gases there are new safety considerations to take into account. In the UK, for example, ethylene oxide is approved in the food industry only for sterilizing raw spices. It is important also to remember that atmospheres with a proportion of oxygen greater than that normally found in air (21%) are potentially hazardous, as oxygen is a reactive element which supports the combustion of almost all materials. Hence materials normally requiring an elevated temperature to burn can readily catch fire in oxygen from the smallest source of ignition. Oils and greases should be removed from pipes, fittings and controls likely to contact oxygen, and oxygen should never be used as a substitute for compressed air.

Table 3.1 summarizes the principal properties of the gases used.

Table 3.1 Summary of gas properties

Gas	Properties
Oxygen	Sustains basic metabolism Prevents anaerobic spoilage

Nitrogen	Chemically inert
	Prevents: oxidation
	        rancidity
	        mould growth
	        insect attack

Carbon dioxide	Inhibits bacterial and mould activity Fat-soluble High concns. can injure produce Not suitable for dairy produce

Applications

Vegetables

Alteration of the atmospheric composition surrounding fresh produce may result in an alteration of its respiration rate. When the oxygen supply is normal, aerobic respiration takes place, but in the absence of oxygen the respiration becomes anaerobic. Lowered concentrations of oxygen will give rise to a mixture of the two processes, and therefore a balance between oxygen and carbon dioxide concentration which is exactly right for the produce and the temperature of storage could achieve a retardation of respiration. In a sealed package, as respiration takes place there will, of course, be a constant change in the composition of the atmosphere as oxygen is consumed. Initially the carbon dioxide concentration rises and the level of oxygen falls, but since the film will always be more permeable to CO_2 than to oxygen, the former will diffuse out through the walls of the package faster than oxygen can diffuse in to replace it. The rate of permeation also depends on the partial pressures of the two gases inside the package and in the surrounding atmosphere, and it is possible that the two rates may in time reach an equilibrium state and a constant atmosphere will be achieved. Tompkins (1967) has, however, shown that this is unlikely to be at a level suitable for preservation, and merely placing fruits and vegetables into sealed packages is more likely to hasten deterioration than to prevent it. The respiration rates of some typical fresh vegetables are given in Table 3.2. Notice the considerable differences between vegetables and the great dependence on temperature.

Because of this, MAP methods for fruit and vegetables need a different approach to the techniques used for meat, fish and baked goods. Suppression of respiration affects the production of ethylene gas, which is for many fruits the principal ripening agent, and also accelerates the gradual deterioration of

Table 3.2 Respiration rates of some typical fresh produce (CO_2 production in mg/kg.h)

Produce	Temperature (°C)				
	0	5	10	15	20
Broccoli	—	—	—	—	425
Calabrese	42	58	105	200	240
Asparagus	28	44	63	105	127
Brussels sprouts	17	30	50	75	90
Lettuce	9	11	17	26	37
Tomato	6	9	15	23	30

both fruits and vegetables. An elevated carbon dioxide concentration has been related to a reduced rate of softening of fruit and vegetables, as well as improved retention of chlorophyll in green vegetables such as asparagus and green beans. As with meat products, there is also an action in retarding microbial spoilage, particularly where related to softening.

There are also hazards in using MAP with fruit. In particular, with the initiation of anaerobic respiration there may be an accumulation of ethylene gas which, as has already been stated, is generally undesirable, and a potential for the growth of *Clostridium botulinum*, which can grow under anaerobic conditions. Also, under low oxygen levels there can be an accumulation of acetaldehyde, ethanol or organic acids, producing discoloration and off-flavours.

Practical examples of the dual technology of preservation carried out at temperatures of 5°C include an increase of shelf life for shredded iceberg lettuce from 10 to 21 days when held in an atmosphere of 2.3% oxygen and 5–6% carbon dioxide, after pre-flushing with an atmosphere of 50% oxygen and 50% carbon dioxide. A similar effect in extending the shelf life of strawberries from 3 to 8 days and raspberries from 1 to 5 days has also been achieved by the same technique. Reported work in Japan has examined the effectiveness of using ethylene absorbers based on potassium permanganate to assist the action of the CO_2. Experimentation to develop better films with a higher permeability to gases suitable for root vegetables and fruits is being carried out at several research establishments in the UK and abroad. Microporous film (van Leer) and high-permeability film (KLF 4 Bunzl) are among the suitable films being studied for fruits and vegetables with high rates of respiration.

Meat packaging

This is an important area for MAP and the factors behind this have been considered earlier. Research has shown that providing the storage temperature of the prepackaged meat is maintained between 1–3°C, then a gas mixture within the package containing 20% CO_2 will inhibit the growth of the

Table 3.3 Shelf-life extensions for MAP meat

Product	Temperature (°C)	Normal package	MAP
Ground Beef	2	2 days	4 days
Liver	2	2 days	6 days
Pork cuts	2	4 days	6–9 days
Beef cuts	2	4 days	10–12 days

principal meat-putrefying organisms, but oxygen must then be present to maintain the bright red colour. Gas mixtures being used for meat packaging include O_2/CO_2, 20/80, and $O_2/CO_2/N_2$, 20/69/11. The nitrogen in the second example is added to slow down the rate of gas absorption mentioned earlier. Films suitable for maintaining this atmosphere for the required time would be, for example, a PVC/PE laminate base tray lidded with PVdC-coated polyester/PE, to which it is possible to add an antifogging coating, as meat packs frequently suffer from this effect. It is also usual to include within the package an absorbent pad because of the likelihood of drip. Table 3.3 gives an indication of the shelf life extension possible with various meat products.

Poultry packaging

Early work on chicken portions stored in aerobic atmospheres enriched with carbon dioxide showed that, within the range 0–25% CO_2, the ratio of shelf life in CO_2-enriched air to that in air alone was a function of the CO_2 concentration. Higher concentrations caused discoloration of the meat, and although there is some argument about whether this limitation exists, in practice concentrations above 25% of CO_2 appear to be advantageous. In the case of skinless turkey breast fillets stored at 1°C in packages with 20% or 30% CO_2, it was found that incorporating 10% or 20% of oxygen in the packages as a possible means of enhancing the colour of the meat during storage led to the rapid development of unpleasant flavours in the cooked meat, an interaction with the CO_2 being indicated. There are certainly differences in the requirement for packaging chicken and turkey, and this also extends in all probability to duck portions as well.

Fish products

As with meat, carbon dioxide inhibits bacterial growth in fish products, but in the case of white fish it tends to be absorbed by the fish tissues, lowering the pH, decreasing the water-holding capacity of the fish, and causing consequent drip. This absorption also causes lowering of the internal package pressure and potential package collapse. Tests have shown that a gas mixture of 40% CO_2, 30% N_2 and 30% O_2 greatly reduces the risk of both package collapse and drip development. A shelf life of up to 9 days can be achieved at 0°C with this mixture, but falls to 5 days at 2°C.

Bakery products

The main problem with bread and flour confectionery is the growth of moulds, and although the use of chemical mould inhibitors can extend the mould-free life of baked goods to more than 50% above normal, there is a risk of developing undesirable odours and flavours. Gas flushing with carbon dioxide is cheaper, and is capable of extending the saleable life of cakes and buns to between three and six months. Other products whose normal shelf life is 48 hours or so can have this extended to between 3 and 4 weeks.

The principles involved in gas-packaging bakery products differ from those of meat and fish in three respects. First, the gas used is 90% CO_2, initially with less than 2% residual oxygen. Secondly, less gas per pack would be used, as the package itself tends to conform more closely to the dimensions of the product, and thirdly, the product is normally stored at ambient temperatures, a great difference from meat and fish, which are always at least chilled. One material that has been found very suitable for these products is a laminate of polyester, polyethylene and polyvinylidene chloride (PET/PE/PVdC). The polyester provides the strength and an aroma/flavour retention barrier, while the PVdC-coated PE provides a gas- and water-vapour barrier and the heat-sealing medium. It is also possible to use metallized or partly metallized films to further increase barrier performance.

Table 3.4 gives some details of typical gas compositions for specific products.

Table 3.4 Typical gas compositions (%) used in MAP

Nitrogen	100	80	70	50	40	30	20	10	—	—
Carbon dioxide	—	20	30	50	60	60	80	20	80	60
Oxygen	—	—	—	—	—	10	—	70	20	40
Thin sliced meats	X	X								
Luncheon meats	X									
Meat-filled pasta			X							
Pizza				X						
Baked goods					X					
Mackerel, herring, sprats						X				
Cheese-filled pasta							X			
Fresh red meat, beef and pork								X		
Fresh red meat and sausage									X	X
Fresh white fish Sole										X

Other products

So far we have considered only a few of the foods which are suitable for this process. Others which have been tried and marketed include pastas, baby foods, nuts, fruit concentrates, cake mixes, preserved meats, cheeses, and snacks. The prime consideration in all of these is, of course, cost and improvement in quality compared with the current product, because unless significant gains can be made either in shelf life or in the perceived and delivered quality of the food, then it is unlikely that MAP will be suitable.

Packaging materials

The material that wraps the product is crucial to the success of MAP. The correct atmosphere at the start will not serve for long if the barrier material allows it to change too rapidly. The properties required in the material are such that few single film materials are completely suitable, and more often than not a multilayer material composed of two or more co-extruded or laminated films is required.

Several factors must be taken into account in determining the combination of properties required for each specific product and market. The more important ones are the following.

(i) The *type of package* to be used (rigid or semi-rigid, lidded tray or flexible film pouch.

(ii) The *barrier properties* needed. In most MAP applications it is desirable to maintain the atmosphere initially introduced into the package for as long a period as possible and to keep the gas ratios unchanged. But it is not always so. Fresh vegetable produce and fruits continue to respire, as we have seen, even when placed in a modified atmosphere, and hence they will produce changes. Under these circumstances the use of a good barrier will be undesirable, and better results will be obtained using materials with poor barrier qualities. Figure 3.4 and 3.5 give typical values for oxygen and water vapour transmission through some materials. It is worth remembering that the values for nitrogen transmission will be about one-third or one-quarter of the oxygen figures, and those for carbon dioxide about 3 to 5 times greater.

(iii) *Mechanical strength*. How resistant to puncture does the material need to be? Are we packaging foods with sharp bones, for example?

(iv) Will the product induce *fogging* and hence will an anti-fog coating be necessary?

(v) *Integrity of sealing*. The better the seal, the more difficulty there will be in opening the pack. The right balance between tightness and security of the closure and the ability to peel back a lidding material must be determined.

There are a dozen or so basic machine systems for handling MAP/vacuum packaging, three of which are essentially used for bulk packs and the others for the smaller retail and consumer packages. Some details are given in Table 3.5.

Limitations of MAP

Despite its many advantages, modified atmosphere packaging does have limitations and attendant costs. There is a requirement to use very high-quality raw materials, strict temperature control, specialized equipment and,

Table 3.5 Basic machine systems for MAP

Type of system	Description	Suppliers
Horizontal FFS, rigid and semi-rigid	Usually consists of a semi-rigid base tray (PVC/PE) lidded with a thinner material. One web is thermoformed and then filled and lidded (Figure 3.2). First developed by Hoechst as the Atmospak' system, Vacuum and gas flush.	Multivac; Kramer & Grébe Mahaffy & Harder Dixie Union
Horizontal and vertical FFS pouches	Usually made from a single film and is totally flexible (Figure 3.3). The system can also wrap a prefilled tray but is capable of only a continuous gas flush. Because of the hot sealing jaws oxygen mixtures cannot be used. Gas flush only.	Rose Forgrove, Aucoutourier, Fuji, Ilapak.
Pre-formed plastic tray or bag	Bags and trays are premade, bags have been used several years for primal meat cuts and cooked meats. Trays are more recent and more usual now. HIPS, HDPE or PET trays. Vacuum and gas flush. Can handle catering sizes.	Dyno, Multivac(space) Maidstone M/c Co, Barnhardt.
FFS composite board/plastic tray	Available in the UK from the Mardon Group. Licensed from SPIC in France and comprises a board blank with thermoformed plastic inserts which are erected inline with a Vacuum/gasflush/lidding machine. These Gemella Seal packages have good graphics.	Mardon/Smiths
Preformed coated pulp trays	The pre-moulded pulp trays have a thermo-formed plastic lining made from a variety of materials and are claimed to be cheaper than plastic trays. Vacuum and gas flush.	Keyes Fibre/ Maidstone Machine Co.
Bag-in-carton	The Hermetet system is probably one of the best known. They are consumer size, for dry powders and granules. Vacuum and gas flush.	Akerlund & Rausing; Bosch
Bag in box	The product is loaded into a barrier or non-barrier bag. It can be in bulk or in conventional EPS trays with film overwrap. The machine inserts 2 snorkels into the top of the bag, draws a vacuum and gas flushes. Has the advantage that conventional unit packs can be produced and then gas-flushed. Only the French Bernhardt m/c operates automatically, the others are manually operated.	CVP, Corr-Vac, Snorkel Vac Bernhardt
Darfresh vacuum skin packaging	Vacuum skin packaging was introduced by W.R. Grace in 1985. The system uses the product as a forming die, and therefore handles different shapes. The top web is 6–8 layers (including HDPE/EVOH and irradiated layer). This web is heated and floated down on to the product to form a skin over it, right up to the edges, and is then welded to all the exposed areas of the bottom web or tray. This is of an easy-peel material, giving good seal strength and easy opening.	W.R. Grace
Cryovac Shrink vacuum bags	Premade bags from PVdC coextruded materials, usually shrinkable with heat. Vacuum only.	W.R. Grace

Bivac vacuum shrink system	2 webs of ionomer film; the upper is heated and shrunk on to the lower web.	Dupont/American Can.
Isopak vacuum skin system	Nylon/Surlyn top web on a polyester/ Surlyn/board backing.	Akerlund & Rausing
Trayvac	Product placed on a flexible web and a rigid tray is brought into contact and the pack is heat-shrunk.	Akerlund & Rausing

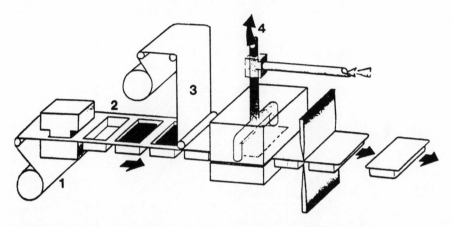

Figure 3.2 Thermoforming packaging machine fed from two film coils. One inner thermo-formable film (1) is formed into a tray (2). The food product is placed in this tray covered by an upper film (3). A vacuum is created in the tray (4), and broken by the gas mixture just before the upper film is sealed.

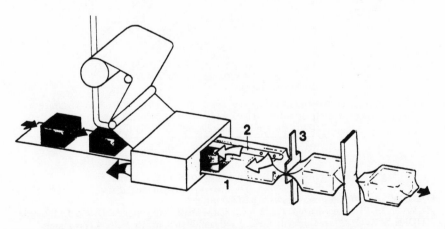

Figure 3.3 Gas flushing. Purging the air from the package is accompanied by continuous gas flushing. The packaging machine provides a film tube (1). Gas is injected into this tube through an injection pipe (2), which extends to a point just before the sealing jaw (3).

of course, the new element in many plants of introducing gas supplies to
operators who are unfamiliar with their use. Probably the greatest problem
lies in temperature control. It has already been emphasized that rates of
deterioration are heavily affected by temperature and the requirements on
gas mixtures can make considerable changes. It is therefore vitally important

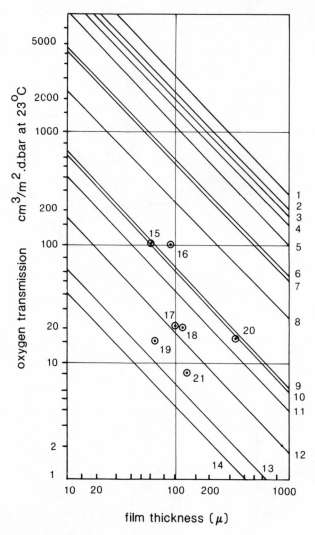

Figure 3.4 Oxygen transmission rates of packaging films.
1, EVA; 2, LDPE; 3, plasticized OVC Stretch film; 4 HIPS; 5, PS (standard); 6, polypropylene;
7, HDPE; 8, PVC (22% plasticized), OPP; 9, polyamide (nylon); 10, VC-VA copolymer;
11, PVC (unplasticized); 12, polyester; 13, nylon 6; 14, PVdC; 15, polyester/PE:12/50;
16, polyester/PE:12/75; 17, polyamide/PE:40/60; 18, polyamide/PP:40/75; 19, polyester/PEX:
15/50; 20, PVC/PE:250/75; 21, polyamide/PEX:60/70.

that everybody from the original packer right through to the retailer is involved in a tightly-controlled chain, and that staff are properly trained in the special needs of the product and the system. The way the products have evolved has also meant that they are usually separately displayed at a much lower packing density than is customary with more conventional foods, and

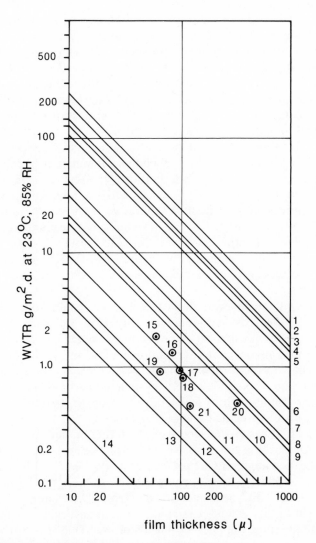

Figure 3.5 Water vapour transmission rates of packaging films.
1, Plasticized OVC stretch film; 2, polyamide (nylon); 3, nylon 6; 4, PS (standard); 5, HIPS; 6, EVA and PVC (20% plasticized); 7, VC-VA copolymer; 8, PVC (plasticized); 9, polyester; 10, LDPE; 11, polypropylene; 12, HDPE; 13, OPP; 14, PVdC; 15, polyester/PE 12/50; 16, polyester/PE 12/75; 17, polyamide/PE 40/60; 18, polyamide/PE 40/75; 19, polyester/PEX 15/50; 20, PVC/PE 250/75; 21, polyamide/PEX 60/70.

this will mean extra costs in the retail store. Equally, quality control is extremely important in the packaging operation and proper testing methods need to be introduced.

Testing of modified atmosphere packs

In the design and production of all types of barrier packaging, the barrier properties of the constituent materials have usually been carefully investigated. With the encouragement of Pira, many companies in the UK have begun to examine measurement of the barrier properties of the complete pack, and this is particularly important where the packaging material is thermoformed into a tray. Equally, with the development of improved methods of testing, we are able to measure the transmission rates of gases and water vapour through complete packs at controlled conditions of temperature and humidity. However, this is not the complete story. Leakage, due to either pinholes in materials or poor seals is much more common than many packers realize. It is therefore extremely important that stringent tests are carried out for leakage, and, bearing in mind that substantial differentials may develop between the inside and outside atmospheres of the pack, these tests must cover both long-term resistance to the development of leaks as well as those detected in the short term. Studies at Pira, for instance, have begun to establish a regime for examining the failure of seals that occurs with the passage of time, so-called 'creep failure', and this is an important factor to be considered.

Equally, it is vital to be aware of the effect of mechanical stress on leakage rates. Many pinholes or leaks only develop after the packs have been subjected to stresses in transport. This can be examined and predicted using controlled conditions on modern vibration tables, and is the best way of ensuring satisfactory performance in the field. Quality control during the machine operation also needs careful examination to establish monitoring procedures for factors such as leakage, peel strength, burst resistance and integrity in transit as well as package properties.

The use of MAP has grown spectacularly during the last five years. More than 100 MAP machine systems have been installed in the UK. The major areas of application at present are for meat, fish, cooked meats and bacon. Its use for bakery products, an area where the UK at present is lagging behind other parts of the EEC, will undoubtedly grow. Developments in barrier technology will further extend possibilities, and in particular the porous and high-permeability films will open new fields of use. Thus, MAP offers many opportunities for food packers, particularly those who can offer premium products. Retailers are becoming increasingly enthusiastic about the types of product that can be offered and are increasing the shelf space available for them. The possibilities in the premium food business are good for the next decade at least.

References and further reading

Anon. (1982) Preservation. *Packaging* **53** (625) 33.
Anon. (1982) Controlled atmosphere packaging. *Packaging News* (September) 65–66.
Anon. (1982) Mould controlled. *Packaging Review* **102** (9) 61.
Anon. (1983) Atmosphere of success. *Packaging Review* **103** (3) 32.
Anon. (1983) Inert gas packaging. *Converter* **20** (8) 16–18.
Blackman, R.R. and Parija, P. (1928) *Proc. Roy. Soc. London B. Biol. Sci.* **103**, 412.
Burg, S.P. and Burg, E.A. (1967) *Plant Physiol.* **42**, 144.
Duckworth, R.D. (1966) Physiology, Chapter 3 in *Fruit and Vegetables*, Pergamon, Oxford, 63–94.
Geeson, J. (1984) Use of plastic films for modified atmospheric packaging of fruit and vegetables, in *Packaging of Fresh Produce*, Pira Packaging Seminar PK/SM/56, 27 November 1984, Session 4.
Henig, Y. (1972) PhD Thesis, Rutgers Univ., New Brunswick, NJ.
Kader, A.A. and Morris, L.E. (1977) in Dewey, D.H. (ed.) *Modified Atmospheres*; an indexed reference list with emphasis on horticultural commodities. Suppl. no. 2, Veg. Crop Ser. 187, Univ. California, Davis.
King, A.A. and Nagel, C.W. (1967) Growth inhibition of a *Pseudomonas* by carbon dioxide. *J. Food Sci.* **35**, 375.
Leeson, R.H. (1984) Developments in the uses of gases for packaging, in *Integrated Food Processing/Packaging Developments*, Pira Packaging Seminar PK/SM/36, 4 April 1984, Session 5.
Leverman, K.W., Nelson, A.T. and Steinburg, M.P. (1966) *J. Food Sci.* **31**, 510.
Magness, J.R. and Diehl, H.C. (1924) *J. Agric. Res.*, 27–33.
Nilhori, I. *Produce Packaging in Japan*. Agricultural Experiment Station, Chiba Prefecture.
Rothwell, T.T. (1986) Modified atmosphere packaging, in *Fresh and Processed Foods*, Pira Packaging Seminar PK/SM/086/A5, 29 April 1986, Session 6.
Tomkins, R.G. (1967) *Food Manuf.* **42**, 34.
Wankier, B.N., Salunkhe, D.E. and Campbell, W.F. (1970) *J. Amer. Soc. Hort. Sci.* **95**, 604.

4 Use of irradiation techniques in food packaging

KIRSTEN NIELSEN

Historical introduction

Ionizing radiation was discovered by Röntgen just before the start of the 20th century. Although at an early stage there was considerable interest in the biological effects, the first patent for the use of radiation to sterilize food did not appear until 1930, and real research on the preservative effects did not commence until the 1940s. In 1943 it was demonstrated that x-rays could increase the shelf life of a hamburger. These researches coincided with the studies on nuclear power and the production of radio isotopes.

The US Atomic Energy Commission (AEC) started a programme on food irradiation in 1950. In the UK at about the same time, studies commenced at the Torry Research Station, Aberdeen, and the Low Temperature Research Station in Cambridge. These programmes were all designed to produce in meat and poultry a reasonably long life at higher quality than that obtained by heat treatment. The AEC programme stopped after a few years, but recommenced in 1960. Each of these programmes demonstrated the effectiveness of radiation sterilization and showed that low doses were desirable from the organoleptic viewpoint, although complete sterility was possible at higher levels.

The US Army Natick Laboratories commenced research on sterilization of food by irradiation in 1962, and by 1964 the FDA had approved the technique for canned bacon, wheat and wheat flour, and for inhibiting sprouting of white potatoes. However, in 1968, because of data felt to show significant adverse effects in animals fed on irradiated feedstuffs, the FDA approval was rescinded on canned bacon and approval not granted on canned ham. Many of the early studies, however, had major deficiencies in design of experiment and the way they were conducted.

Research and development in this area has continued in the USA and the UK food industries, but after the initial interest, potential problems of health were raised, and the commercial possibilities tended to recede so that interest declined. Research was also started in other countries: in Canada and Japan (1956), the USSR and Argentina (1957), Poland (1958), India (1959) and Israel (1960). By 1968, 76 countries had food irradiation programmes. In Europe, both the EEC and OECD contributed to the developments. In the

Table 4.1 History of food irradiation

1895	X-rays discovered by Röntgen
1930	Patent for radiation sterilization of food
1940s	Nuclear power and production of radioisotopes. Accelerators for generating electron beams.
1950s	Research commenced by US Atomic Energy Commission, by Torry Research Station and LTRS in UK.
1960	First full-scale irradiation plant in UK built at UKAEA, Harwell.
1962	US Army Natick Laboratories start research
1964	FDA approve limited use. UK government receive advisory report.
1967	UK (Control of Irradiation) Regulations
1970s	International Study Project, Karlsruhe, FRG.
1981	Joint FAO/WHO/IAEA report on the wholesomeness of irradiated food. FAO-WHO Codex Alimentarius Commission produce general standard and code of practice for food irradiation
1982	UK form Advisory Committee (ACINF)
1986	US FDA rules: Fresh fruit and vegetables 1kGy Herbs and spices 30kGy All packs to be labelled specially and carry an international vignette UK ACINF produce their report

1970s an international project was started at Karlsruhe in the Federal Republic of Germany. 1981 provided a report from a joint project financed by FAO/WHO/IAEA — the JECFI report. The Codex Alimentarius Commission also produced a general standard for such foods and a Code of Practice for food irradiation facilities (see also Table 4.1) Several countries since 1984 have attempted to rationalize possibilities and to standardize practices, so it can be seen that this is a growing market.

The technology

Electrons and photons, whether they come from an accelerator or a radioactive source, such as cobalt-60, constitute a form of energy just like visible light. In fact they form part of the spectrum of energy which extends from radio waves through the visible to high-energy electrons (Figure 4.1). This energy is capable of inactivating micro- and other organisms. The wavelength of such radiation is, however, much shorter than that of the visible light spectrum. Because of this short wavelength and their higher energy, these radiations have a high penetration power. Hence, one of the characteristic advantages of the process is that the product can be irradiated when it has been placed within its container, so that recontamination after processing is prevented. Moreover, it is not possible for the product itself to become radioactive, and there are no residues of any kind left by the process; once treated, foods are ready for use or consumption, whether they are fresh or processed before packaging.

The inactivation of food spoilage micro-organisms is brought about by changes in the DNA molecules in the living cells. Because of its size and other

c

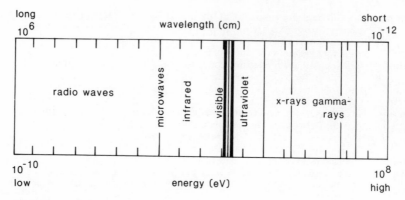

Figure 4.1 The electromagnetic spectrum.

properties, the DNA molecule is far more sensitive to the radiation than the molecules of the food undergoing processing. As a result, bacteria, moulds and yeasts are killed long before any undue changes can take place in the flavour of the food itself. In contrast to heat processing (canning, bottling and pasteurization), the product does not change, since its temperature is not raised significantly and no cooking occurs. This can be a major advantage where heat-sensitive products are involved.

The effect of irradiation depends on the exposure dose, which in turn is determined by the residence time in the irradiation chamber. Since the power of any one source is constant, this is the only variable in processing, so that the process is most reliable and reproducible.

The relationship between the exposure dose and the concentration, C, of the reaction products in mg/kg is given by the equation

$$C = 1.04 \cdot G \cdot M \cdot D \times 10^{-4}$$

where G = number of reacted molecules per absorbed energy unit of 100eV.
        M = molecular weight
        D = dose in Gy.

Techniques

There are two possible methods of irradiation for foods at present: low-dose gamma-radiation, and low-dose irradiation by electron beam.

Low-dose gamma-irradiation

A plant using low-dose gamma-radiation comprises a radiation source, protective shielding, conveyor systems both inside and outside the radiation chamber, together with control and safety equipment. The heart of the system is the cobalt unit, which uses cobalt-60 having a specific radioactivity of more than 100 curies per gram. This material is assembled in the form of tiny cylinders to form rods 450mm long. These rods are fully screened by hermetically-welded double walls of stainless steel. A rod of cobalt-60 usually contains 10 000 curies. A vessel filled with water, as a shield against radiation, serves as storage accommodation for the cobalt unit.

The walls of the irradiation chamber and spaces are made of 1.8m-thick concrete to ensure a safe shield for the surrounding areas, which include the control room and the production hall. Safety equipment is also housed in the control room. To obtain optimum utilization of the plant, a system of conveyors is selected which will allow the products to be continuously supplied and discharged, 24 hours a day for seven days a week. The process is supervised by a telemetric monitoring system, and bulk goods can be handled in pallet loads in many instances. Products for medical and surgical application were the first to be sterilized by irradiation; commercial use for foods is still experimental.

Low-dose electron-beam radiation

A typical accelerator consists of an evacuated tube to the ends of which an electric potential difference is applied. At one end we have an ion source injecting charged particles into the tube. These arrive at the other end with higher energy. They then impinge on the target (the food) and their energy is transferred to it as it passes by on the conveyor. Since the penetration of the beam is low, the treatment is repeated from the other side of the product. The exposure time is of the order of minutes, according to the nature and quantity of the product.

Applications

The potential range of applications is very broad. The possibility of in-activating spoilage organisms without significant change to the food and without using any chemical methods of preservation offers many potential benefits for a variety of products.

Inactivation of pathogens

The incidence of diseases caused by micro-organisms has increased alarm-ingly during recent years. Food poisoning from *Salmonella*, *Staphylococcus* and *Botulinum* as well as parasitic infections have been increasingly reported. Such outbreaks are caused by eating infected meat, meat products, poultry, fish, seafoods, frogs' legs, egg products, cereals which have mildewed and nuts which have been kept too long. Thermal treatment is not possible for such products because they are required in the raw (uncooked) state. But low doses of radiation are sufficient to inactivate the pathogens, most of which are more sensitive to irradiation than the normal spoilage flora. Mossel (1977) has estimated that a 400 krad dose will reduce the amount of salmonellae by some six log cycles.

The food can be irradiated after having been hermetically packaged, so that reinfection is precluded. An attendant advantage is the partial elimination of psychrotrophic rod-like bacteria which cause decay, and as a result the durability of refrigerated products can be extended. Heat treatment of quick-frozen products is an unattractive concept, both from the energy viewpoint and in terms of texture of the food.

Disinfection of cereals and flour products

This treatment will destroy any insects, larvae and eggs, and is particularly useful in Third World countries, where losses of cereal foods to insects are considerable. Tilton and Burditt (in Josephson and Peterson, 1983) have discussed the subject in a recent review, according to which the large number of insect species present in grains makes it imperative that the radiation dose is large enough to kill or sterilize the most resistant species present. They also found that age, sex, food and temperature can affect the sensitivity of any species. In general females are more sensitive than males, and the Lepidoptera are more resistant to gamma-radiation than the Coleoptera. A combination of gamma-radiation and microwaves or infrared treatment has a synergistic effect, the combination being more effective than either alone.

Sterilization of packaging materials for the food industries

This can particularly useful where aseptic packaging is involved (see Chapter 2).

Disinfection of herbs and spices

Many spices are highly contaminated with bacteria and fungi. Coriander, ginger, caraway, black pepper and turmeric typically have aerobic plate counts of 80 million bacteria and 100–10 000 moulds per gram of spice. While it is possible to fumigate with ethylene or propylene oxide, there are undesirable reactions and fumigation can take hours. Irradiation can reduce microbial contamination to below 1000 per gram using a radiation dose of 0.5–0.8 Mrad (Farkis in Josephson and Peterson, 1983). Moreover, radiation treatments can be automated and applied to prepacked materials.

Extending the life of perishable foods

Irradiation can eliminate the major part of the micro-organisms responsible for spoilage. Low-dose radiation preservation of seafoods, for example, does not degrade the nutritional quality any more than heat preservation, and would stabilize supplies and expand the potential markets. The life of fruits such as strawberries can be extended by a factor of almost two if the growth of mildew and moulds is inhibited by irradiation.

Inhibiting sprouting

By using very low radiation doses, tubers and bulbs such as potatoes and onions are prevented from germinating, and less satisfactory treatments with chemical anti-germinants are obviated.

Decontamination of soils

The undesired growth of fungi, insects, weeds and other organisms leading to decay can be prevented.

Sterilizing food for prolonged storage

A combination of short heat treatment sufficient to inactivate enzymes, and irradiation to inactivate micro-organisms, permits the use of food preservatives to be restricted to the very minimum. Because of the poor heat penetration in solid products of low moisture content, sterilization by heat cannot be implemented without destroying the essential nutrients. For products like meat, the combination of heat treatment and irradiation can give a long-life product without the use of preservatives or refrigerated storage. As examples of its use, we may cite the production of foods for space travel and for building emergency stores.

Legislation

The commercial use of irradiation in preserving food has always been treated with caution. At present a number of countries permit its limited use. In the USA, irradiation is classified as a food additive and regulated by the FDA (Food and Drug Administration). The relevant documents are Title 21, paragraphs 179.22 and 179.24, which deal with low-dose gamma and electron-beam radiation for food treatment respectively. Paragraph 179.22 permits the treatment of certain foods by gamma-radiation under the following conditions:

(i)   The source consists of sealed units of cobalt-60 or caesium-137
(ii) The radiation is used or is intended for use in a single treatment for the foods listed below under the specified limits

Food	Maximum absorbed dose	Use
Wheat	20–50 000rad	Control of insect infestations
Wheat flour from unirradiated wheat	20–50 000rad	Control of insect infestations
White potatoes	5–15 000rad	Inhibition of sprouting

(iii) The label shall bear in addition to all other statutory requirements the following statements: *Treated with ionizing radiation* or *Treated with gamma radiation* on all retail packages and *Treated with ionizing radiation — do not iradiate again* or *Treated with gamma-radiation — do not irradiate again* on wholesale packages and on invoices or bills of lading of bulk shipments.

Paragraph 179.24 deals similarly with low-dose electron-beam treatment as follow:

(i)   The radiation source consists of an electron accelerator producing a beam of electrons at energy levels not to exceed 5 million electron volts (Mev)

(ii)   The radiation is used or intended for use in a single treatment for the listed foods as follows:

Food	Limitations	Use
Wheat and wheat flour from unirradiated wheat	Absorbed dose: 20–50 000rads. Maximum thickness of food 0.6 cm/Mev of electron energy under single beam irradiation or 1.4 cm/Mev with crossfiring beams. Maximum flow 10 tonnes per hour per kilowatt for single beam or 14 t/h/kW with crossfiring beams.	Control of insect infestations

(iii) A permanent record of the radiation intensity and power used shall be made with recorders coupled to the accelerator, and these shall be retained for FDA inspection for a period of one year. Such records shall completely identify the food treated.

(iv) The same labelling requirements as with gamma-radiation apply for electron-beam treatment.

In the United Kingdom at present the use of ionizing radiation on consumer food is prohibited, and also the import and sale of food which has been so treated. However, a recent report from the UK Advisory Committee on Irradiated and Novel Foods (1986) states:

> We are satisfied...that ionizing radiation up to an overall average dose of 10 kilogray (kGy) correctly applied, provides an efficacious food preservation treatment which will not lead to a significant change in the natural radioactivity of the food or prejudice the safety or wholesomeness of the food.

Similar restrictions are applied in the Federal Republic of Germany. The treatment is being used under carefully controlled conditions in several countries including France, Belgium, The Netherlands, Japan and Canada. The World Health Organisation and the UN Food and Agriculture Organisation's Joint Expert Committee have stated that irradiation up to an average

Figure 4.2 The internationally-recognized vignette for irradiated foods.

dose of 10kGy presents no toxicological hazard and introduces no special nutritional or microbiological problems. Specifically on the nutritional aspects, this Committee concludes that the nutritional consequences of irradiating food are no different, qualitatively or quantitatively, from the nutritional consequences of currently accepted methods of processing and preparing foods.

In many of the countries permitting the sale of such foods the packaging must be made identifiable with an internationally recognized vignette (Figure 4.2).

Standardization

FAO and WHO have had a joint standardization programme since 1962, and this includes the irradiation of foods. The work is conducted under the aegis of the Codex Alimentarius Commission. To date the studies have produced two draft proposals: a revised draft recommended international General Standard for irradiated foods, and a revised draft recommended international Code of Practice for the operation of irradiation facilities used for the irradiation of foods.

The first draft proposal is applicable only if the food is *processed* by irradiation. It does not apply if the foods are exposed to radiation doses imparted by measuring instruments used for inspection purposes. It deals with questions related to absorbed dose, facilities and control of the process, wholesomeness of the foods, technological requirements, packaging and food quality requirements, re-irradiation and labelling.

The second proposal for a draft Code of Practice, refers to the operation of irradiation facilities based on the use of either a radionuclide source (cobalt-60 or caesium-137) or x-rays and electrons generated from machine sources. The facility may be one of two designs, either 'continuous' or 'batch' type. Control of the process in all facilities involves the use of accepted methods of measuring the absorbed radiation dose and the monitoring of the physical

parameters of the process. The operation of these must comply with the
Codex recommendations on food hygiene. The standard deals with questions
related to irradiation plants, good radiation processing practice and product
and inventory control.

Effects on packaging

One of the advantages of the process is the fact that many foods can be
treated after they have been packaged. This assists in keeping the food sterile
until consumed. It is essential therefore that the irradiation process should
have no significant effect on the properties of the packaging materials used.

Various studies have been undertaken to measure any such changes which
might occur in relevant material properties consequent upon the irradiation
process. This is particularly important for many plastic materials and the
additives used in processing them. Possible migration of additives, oligomers
and low-molecular-weight portions of the polymer is one of the more
important aspects, and basic studies have been made (Payne *et al.*, 1962;
Timmermann, 1963). The effect of several levels of radiation dose on the
structure of an LDPE film was measured by means of infrared spectroscopy.
Irradiation was found to change the infrared spectrum but not the UV
spectrum of the material. Similar studies were later made on polypropylene
film (Varsanyi, 1962) when statistically important changes were observed in
both the IR and UV spectra.

Figge (1977) studied the effects of irradiation of plastic film in respect of
migration into the fatty food simulant, HB307. Results showed that when
irradiated with 2.5 Mrad, both polyethylene and polypropylene films undergo
a change which dramatically reduces the migration of some additives, while
PVC was almost unaffected at this level. Nielsen (1985) has shown that
radiation doses up to 30kGy have no effect on the tensile strength, the
water vapour transmission rate, and the rate of transmission of oxygen of
polyethylene and polyamide/polyethylene laminate bags.

Studies for the FDA in the USA have also been made, and that body has
concluded that the packaging materials listed below may be safely subjected
to irradiation at the following levels.

Up to 1 Mrad	Up to 6 Mrad
Coated cellulose film	Vegetable parchment
Glassine paper	PE films
Waxed paperboard	PET
Kraft paper	Nylon 6
PET	PVdC–PVAc copolymer
Polystyrene	Acrylonitrile copolymer
PVdC film	
Nylon 11	

Summarizing, we may conclude that food technologists have developed an important process which can remove the causes of both deterioration and poisoning while leaving the food in as near a natural state as possible in appearance, texture, taste and nutritional value.

References

FDA Code of Federal Regulations, **21** (1985).

Figge, K. *et al.* (1977) *Dtsch. Lebensmittel Rdsch.* **73** (7) 205.

Josephson, E.S. and Petersen, M.M. (eds.) (1983) *Preservation of Food by Ionising Radiation.* Vol. III, CRC Press, Baton Rouge, Florida.

Mossel, D.A.A. (1977) The elimination of enteric bacterial pathogens from food...with reference to *Salmonella* radioexcitation. *J. Food Quality* **1**, 85.

Nielsen, K. *et al.* (1985) Electronbestraling af emballage-materiale til krydderier, ETI.

Payne, G.O. *et al.*, US Army Contr. NO DA-19-129-AMC-162.

Timmermann, R. (1963) Radiation Dynamics Inc., Westbury, Long Island USA.

Varsanyi, I., *et al.* (1962) *J. Acta Alimentaria.* **1** (1), 5.

5 Shelf-life prediction

DENNIS J. HINE

The term 'shelf life' is generally understood to be the duration of that period, between packing a product and using it, for which the quality of the product remains acceptable to the product user. An attempt to predict this period from data on the product, the pack and the distribution and storage conditions is appropriate where alternative packaging materials are available which make positive yet potentially differing contributions to the extension of the usable life of the contents.

Shelf-life prediction would not be appropriate for canned goods or products vacuum-packed in glass jars, where it must be assumed protection from external climate is total and deterioration in product quality is through interaction between product and container or through defects in pack sealing, or because of inherent instability of the product itself. Prediction is also inappropriate in frozen-food packs where low temperature is the major preservation means and the pack has the minor role of prevention of dehydration.

It is, however, appropriate for pharmaceutical products in lidded thermo-formed blisters where moisture transmission can reduce the potency of the tablets or capsules. Oxygen ingress can reduce the quality of bag-in-box wine, hence the concern for the permeability of bag and tap and their effect on shelf life. Modified atmospheres rich in carbon dioxide can extend the successful storage of vegetable produce, meat and fish. Correct choice of barrier can slow CO_2 loss and give the right oxygen and relative humidity conditions in the pack so that product deterioration is retarded.

A rather different aspect of shelf life is the time before migration of components from a plastic packaging material into a foodstuff reaches the maximum legal limit for that component, or the food is made unpalatable.

Incentives for shelf-life prediction studies

Shelf life is above all a marketing concept of economic importance in package development. Extension of the storage life of a fresh product by a few days may enable deliveries to retailers to be reduced from two to one a week, with a consequent saving in transport costs. The right shelf life may allow the year-round sale of a seasonal product or the year-round production of a product with a seasonal sales pattern.

Shelf-life prediction can be said to be a means of reduction of storage trials to those combinations of package and product that are likely to give the desired shelf life set by consideration of the sales and distribution pattern seen for the product. Economies can come directly through the reduction in trials, the choice of the most cost-effective pack and the reduced risk of failure of production packs to afford adequate protection. The prediction process involves the properties of the product, and can show that changes in product formulation are desirable to reduce demands on the protection expected of the pack and thus reduce packing costs.

Factors in the shelf-life prediction process

The ability to predict shelf life depends on the availability of data on

(i)   The mechanism of deterioration of the product
(ii)   The agents responsible for control of the rate of deterioration
(iii)   The quality of product in the pack
(iv)   The desirable shape and size of the package
(v)   The 'quality' of the product when packed
(vi)   The minimum acceptable quality of the product
(vii)   The climatic variations likely to be encountered during distribution and storage
(viii) The mechanical hazards of distribution and storage that may affect the integrity of packs
(ix)   The distribution unit, whether this be an individual pack or a collation of such packs in a transit pack
(x)   The barrier properties of packaging materials against the agents causing product deterioration
(xi)   The influence of conversion of packaging materials into packs on the barrier properties
(xii)   The significance and distribution of defects in barrier performances of production packs.

The task of a shelf-life prediction technique is to take this data, together with marketing requirements, to give an estimate of the time before the product is unacceptable. The definition of what is acceptable is crucial to this process. Setting this limit follows from an understanding of item (i).

The mechanisms of deterioration

Water vapour

Many of the deteriorations that occur in packed goods are associated with gain or loss of water vapour. Moisture exchange with the surroundings can cause physical changes, alter flavour, reduce the effectiveness of drugs or promote mould and bacterial growth.

The importance of water vapour in the deterioration of many foodstuffs is demonstrated in a tabular presentation of specific deterioration indices for classes of foods given by Paine and Paine (1983). It is shown to be most important for baked goods, processed fish, leaf vegetables, cereals and cereal products and confectionery.

More specific changes in food properties are discussed by Oswin (1976), who tabulates the various roles of water in foods. These include its role as solvent; as a medium in which reactions can proceed; as a reactant in its own right; and a means of making structural changes in texture, viscosity and other rheological properties. Oswin outlines the relative humidity (water activity) ranges where characteristic deteriorations occur. The general pattern is for oxidation to be favoured at water activities below 0.5 (50 r.h.), browning over the mid-range of 0.3 to 0.75, while mould growth and then bacterial infection are accelerated by values of water activity over 0.7 and 0.85 respectively.

This subject is treated in more detail by Karel (1975), who stresses the importance of water activity for the growth of micro-organisms. The water activity ranges which foodstuffs characteristically exhibit are charted, together with the minimum water activities at which specific micro-organisms start growth. The consequence of this link is that packaging may have two roles, depending on the circumstances. In a baked product such as a pie where the interior may be moister than the surface crust, the pack should allow the surface to lose water so that the water activity there is below that for mould growth. The function of the packaging material is then hygienic, to prevent contamination and soiling. For a dry baked biscuit, the role of the packaging material is reversed. The barrier is required to prevent moisture pick-up which can lead to a musty taste associated with incipient mould growth. However, before this state is reached, the biscuits will probably be judged unpalatable through loss of their initial crispness. This is one of a variety of physical changes. Moisture loss can cause wilting of leaf vegetables and surface hardening of cakes. A rise in moisture content can cause expansion and disintegration of pharmaceutical tablets or retard the rate of solution of a packed powdered beverage. Colour changes may occur in fresh meats, although these are also determined by oxidation reactions.

Oxidation

Karel (1975) comments that the oxidation of cured meat pigment to an undesirable brown colour is approximately proportional in rate to the partial pressure of oxygen. However, for fresh meats, the rate of discoloration at first rises, then peaks at a relatively low oxygen pressure and thereafter decreases.

Many other reactions with oxygen have undesirable consequences. Vitamins are affected and lose their potency. Ground coffee suffers in both aroma and taste by exposure to air, hence the use of vacuum or carbon-dioxide-flushed packs for this product. The respiration of fresh fruits and vegetables after harvesting makes high demands on available oxygen. Respiration and thus ripening can be retarded by reduction in oxygen partial pressure. However, an increase in spoilage rate may occur at very low oxygen pressures due to a change from aerobic to anaerobic respiration.

The best-known effect of oxygen on deterioration is the promotion of the rancidity of fats. Again this is a complex phenomenon. Moisture content may affect the rate, and there can be an induction period after exposure of the oil or fat to oxygen before rancidity is noticed. Diffusion into the fat can reduce the rate of development of rancidity, making thin fat layers most vulnerable. This can occur in potato crisps, for example. Work on potato crisps has shown the role of anti-oxidants in the coating oil. A point that should be noted here is that the packaging material may have little effect as a gas barrier on the progress of oxidation as there is adequate oxygen in the air packed around the crisps. The packaging material, if a light barrier, may reduce the rate of rancidity development, as it decreases the catalytic effect of light on the oxidation reaction.

Light

The intensity of light and its wavelength are factors in considering degradation caused by exposure. Short, ultraviolet wavelengths are often more dangerous than visible or infrared ones. Light can produce visual changes. Some pharmaceutical tablets may darken, and pigment colours may fade whether they be natural or added components of foodstuffs. Vitamins such as riboflavin and ascorbic acid may lose their activity through light exposure (Anon, 1979).

Volatiles

Many of the degradations caused by moisture, oxygen or light exposure will be a combination of a physical change in texture or appearance and a change in taste of a foodstuff. Taste and aroma can also depend on essential oils and spices in the foodstuff. If these permeate through the package walls, then acceptability by the consumer will be impaired. The reverse may also occur; taint can be caused by diffusion of, for example, say the smell of diesel oil; the perfume from a detergent or the aroma from the oil of citrus fruits, through the packaging material and absorption by the pack's contents.

Cosmetics may suffer in a similar way by loss of perfume through the container. Perfumes are complex mixtures in which loss of the more volatile constituents can occur more rapidly than other components. Thus the balance of the residual perfume may be lost, and its acceptance by the wearer impaired.

Temperature

Although the thermal conductivity and reflectivity of barriers may retard or modify the influence of external temperature changes on a packed product, the effect is not in the same category as protection from degradation initiated by moisture or oxygen, for example.

The principal influence of temperature changes is on the rate or nature of degradations by other causes. Temperature may have a double effect in that it modifies the barrier properties of plastics and also affects, for instance, the rate of oxidation of fats. Low temperatures have a special influence, as not only does the water/ice transition reduce water-vapour pressure, but crystallinity can be induced in plastics with a consequent reduction in permeability.

Elevated temperatures, by increase in mobility and decrease in viscosity, can cause fat separation and enhanced migration into packaging materials.

The humidity inside a sealed package where there is water absorbed in the product will rise if the temperature rises. In these circumstances liquid water can condense on, for example, the cooler, shaded inside of a package left in sunlight. This liquid water may provide the conditions locally for mould growth or metal corrosion.

If the package is not properly sealed, the package can, in effect, breathe as the daily rise and fall in temperature expands and contracts the air in the pack and forces exchange with the external atmosphere. The effect of holes in packs is considered later when dealing with shelf-life distributions for commercially produced packages. Firstly, however, shelf-life prediction for a perfect pack has to be analysed.

Basic shelf life prediction data

The foregoing brief account of degradation mechanisms is intended to illustrate the ways in which the quality of a packed product can deteriorate. In a package storage trial at temperate conditions of 25°C, 75% r.h., or the accelerating tropical conditions of 38°C, 90% r.h., it is usual to follow the deterioration of food products using taste panels, visual observations or chemical analysis of (say) riboflavin content. In this way a direct measurement of a maximum storage time until an unacceptable loss of product quality occurs can be determined. In other words, a direct determination of the shelf life of the product–pack combination under defined climatic conditions is made. The use of tropical, compared with temperate, storage can shorten shelf life by, theoretically, a factor of 2.4, provided that the rise in temperature does not alter the type of deterioration by water vapour.

Prediction of shelf life implies a mathematical process which needs deterioration to be quantified and related to permeation properties of the barrier material. For example, if biscuit crispness assessed by taste has reached an unacceptable low level, the corresponding biscuit moisture content has to be measured. Prediction with other packs can then use moisture gain as an indicator of deterioration. In a similar way, rancidity can be related to total oxygen uptake and the prediction equation will feature oxygen permeability. Continuing with the biscuit example, it is possible to use a mathematical prediction process making use of moisture content, taking the absorption process up to a moisture-content limit equivalent to the point of acceptability for crispness. The essential information is then the quantity of water that has to permeate the barrier pack to reach this critical limit.

The moisture absorption isotherm

The water content of a foodstuff or other moisture-sensitive product, and the relative humidity (or water activity) with which it is in equilibrium are linked by a characteristic curve for the product. If the product is placed in an atmosphere with which it is not in equilibrium, its moisture content will alter to bring it to equilibrium. The final moisture content usually differs for a given relative humidity, depending on whether the product has lost or gained moisture to reach equilibrium.

The experimental technique to obtain the water isotherm has been standardized following the COST 90 project of the European Cooperation in Scientific and Technical Research (Jowett, 1984). Saturated salt solutions are used in temperature-controlled enclosures to provide air of known relative humidity. Quantities of the product are exposed in these enclosures until weight equilibrium is established. A standard reference material, microcrystalline cellulose (MCC), is used to prove the reliability of the absorption isotherm determination. Given a satisfactory match to the master MCC isotherm, the actual product can be evaluated with assurance.

The absorption isotherm can also be studied by inverse gas chromatography. In this method, the product, which could be pulverized biscuit or soluble coffee, is made the stationary phase in a chromatograph column (Helen, 1985). The column is temperature-controlled and a stream of helium passed through it. A fixed concentration of water vapour is now introduced into the helium stream. When the product in the column is saturated at this concentration, the detector in the helium output from the column will indicate the presence of water vapour. The quantity of water absorbed is calculated from the flow rate, water vapour concentration in the helium stream and the time between the start of vapour introduction into the stream and the response of the detector to the presence of water vapour at the output of the column. Repeating this measurement at other water-vapour concentrations allows the absorption isotherm to be plotted. The water absorption isotherm for falling moisture content can be determined by the reverse process of substitution of a dry helium stream for a moist one, making the product desorb. These techniques are described by Apostolopoulos and

Gilbert (1983) in a paper dealing with the water sorption of coffee or solubles. The absorption isotherms are presented for temperatures of 25–45°C. Generally, at constant water vapour pressure, the moisture content of the product in equilibrium rises with fall in temperature (Gilbert, 1984).

Oxygen absorption

Although water-vapour absorption can be measured by gravimetric or gas chromatographic means, the process of oxidation presents a greater problem, as a smaller mass of gas is involved. Inverse gas chromatography has been used to quantify the amounts of oxygen reacting. Gas streams of a range of oxygen concentrations are supplied to a column packed with the product to be studied. The time between the start of oxygen introduction and the detection of oxygen at the exit from the column is a key measurement. The absorption isotherm for oxygen is derived from these experiments with differing input oxygen concentrations.

Some oxidation reactions vary in rate, depending on moisture content and the illumination conditions. Water vapour can be introduced along with oxygen, and the light intensity can be controlled if the column has transparent quartz walls.

There are two situations in packaging where oxygen ingress is of concern. One of these is the vacuum pack where there is an absolute pressure difference across the walls of the pack; the other is an inert gas flushed pack, where a partial pressure difference of oxygen exists across the barrier. The initial state can be determined by the residual absorbed oxygen in the product as packed, in either situation.

The storage climate

Deterioration in quality of a product in a barrier pack is controlled by the exchange of vapours and gases through the barrier with the surrounding atmosphere. Moisture vapour and oxygen are constituents of general concern, but it may be necessary to consider the exchange of volatile aromas from the product or to the product from the surroundings. As well as oxygen, transmission of nitrogen and carbon dioxide may have to be taken into account in packages where the concentration of these gases in the pack has been artificially raised to give a preservative action.

In the main, the partial-pressure difference of the vapour or gas across the barrier will control permeation. Exchange can also occur due to pinholes in the material or channels in seals and closures.

The gaseous composition of the Earth's atmosphere (Table 5.1) is essentially constant at sea level. Thus, as was stated earlier, the partial-pressure difference of gases across the barrier which causes permeation depends on the internal atmosphere of the pack when this was closed around the product.

Table 5.1 Composition of the atmosphere (percentage by mass)

Nitrogen	75.60
Oxygen	23.05
Argon	1.30
Carbon dioxide	0.047

Traces of neon, helium, krypton, hydrogen and xenon

However, the partial pressure of water vapour in the atmosphere varies continuously. The general pattern is for temperature to rise and for relative humidity to fall towards midday, and thereafter to reverse, temperature falling and humidity rising through afternoon and night. This pattern is modified by for example rainfall and by the seasons. The magnitudes of these variations in external climate are detailed in BS 4672: 1971, but the storage climates in buildings are only broadly related to the external climate of the weather station report. This is well documented in a survey by BFMIRA and Pira (Cairns *et al.*, 1971). Climatic variations in temperature and humidity can differ as much between different building constructions as between seasons on one site. Solar heating can be a factor during lorry and van transport. The mass of packs being stored or transported at a time can diminish fluctuations in the microclimate at the barrier. This effect is in part due to the thermal inertia of packages in bulk and in part by the shielding effect of packs in close proximity.

This concern for the climate, temperature and humidity levels and variations that packs might encounter during storage and distribution has to be allied with the level of barrier of the pack. The better the barrier, the less significant become short-duration humidity excursions of (say) 2–3 hours. Before a change in humidity can influence vapour transmission rate, the gradient of water-vapour concentration from the exterior to the interior of the packaging material has to be altered. A 10% r.h. change in external humidity would have to be sustained for 6h when the barrier has a permeability about $5g/m^2$ day before the internal humidity of a pack is noticeably affected. Table 5.2 shows the influence of season and type of premises on quarterly mean temperatures and humidities.

Table 5.2 UK climate in typical premises (after Cairns *et al.*, 1971)

Period	Supermarkets		Heated warehouses		Unheated buildings	
	°C	r.h. (%)	°C	r.h. (%)	°C	r.h. (%)
March–May	17.3	48.3	15.9	53.4	11.1	67.4
June–Aug	21.1	53.9	20.8	66.6	17.1	66.5
Sept–Nov	18.2	55.9	17.0	60.8	12.7	75.3
Dec–Feb	14.5	48.0	12.7	55.3	4.8	79.2

Values for package permeability

Water vapour

Shelf-life prediction has been discussed so far in terms of understanding the degradation machanisms, the agents responsible for these degradations and the climates in which products are stored and distributed. Control of shelf life can be sought through control of the temperature of distribution or the barrier afforded by the pack to moisture vapour, oxygen and other gases, volatiles and light. Control by incorporation of, for example, chemical antioxidants or fungicides in the product or pack has not been considered, although in certain circumstances current regulations may allow these techniques to be used.

A wealth of information is available from packaging material suppliers' data sheets and literature sources on the permeability of packaging materials to water vapour, oxygen and to a lesser extent, carbon dioxide and volatile aromas. This information on sheet materials needs examination for the extent to which it can be used in the context of production packages and the climates of distribution.

Above a water vapour transmission rate of $1g/m^2$ day, permeability can be measured by gravimetric methods, specified in BS 3177:1959; BS 2782: Part 5: Methods 513 A and B: 1970; ISO R 1195: 1970; ASTM E96–66: 1972; and DIN 53122. The permeability is usually measured either at temperate conditions (25°C, 75% r.h.) or tropical conditions (38°C, 90% r.h.), or, as an alternative, 25°C, 85% r.h. is sometimes used.

The high humidity is maintained on one side of the barrier by a conditioning chamber, the other side of the barrier being kept at approximately zero relative humidity by use of anhydrous calcium chloride.

A number of other commercial methods are available (e.g. Figure 5.1). These are not covered by standards, although they can be shown to give comparable values. They have the advantage that values as low as $0.01g/m^2$ day can be measured. The moisture which permeates the barrier may be detected by the absorption of infrared radiation, or absorbed on the surface of an electrode coated with phosphorus pentoxide. The absorbed water is decomposed by electrolysis and the quantity of water deduced from the electrolytic current. These water-vapour transmission rate techniques are described by Paine and Paine (1983) and Sweeting (1971).

These methods give essentially the permeability of undamaged sheet materials. Method BS 3177 describes a creasing technique that can be applied to test pieces to assess the possible increase in permeability when sheets are folded, as in making wrappers on packs. Heat sealing effects can be included by making pouches containing desiccant such us silica gel or anhydrous calcium chloride and using the gravimetric technique if the packaging material is suitable.

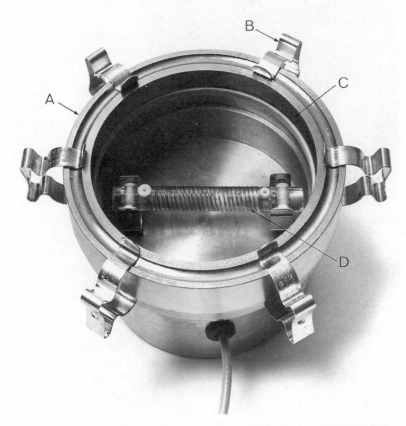

Figure 5.1 The TNO/Pira cell. *A*, ring; *B*, clip; *C*, film to be tested secured with a gasket under the ring; *D*, element.

Obviously this direct weighing method can be extended to other packages, such as thermoforms and lidded or sealed jars. In these studies, it is essential that as well as weighing packs containing desiccant, control packs containing an equivalent mass of inert contents such as glass beads are stored and weighed under the same conditions. In this way, absorption of water by the pack itself and weighing errors can be detected. This is of particular importance when the permeability of the pack itself is low.

Determinations of pack or packaging material permeabilities by weighing may take from one to four weeks, depending on the level. Tropical conditions may be called 'accelerated' storage. The saturation vapour pressure of water increases with temperature as shown in Table 5.3. Allowing for the differences in relative humidity as well as temperature, the ratio of vapour

Table 5.3 Saturation vapour pressure of water at various temperatures

Temperature (°C)	SVP (mm Hg)	Temperature (°C)	SVP (mm Hg)
−20	0.79	20	17.5
−10	1.97	30	31.8
0	4.58	40	55.3
10	9.20	50	92.5

pressure at tropical to that at temperate conditions is 2.4. This acceleration factor should be noted only as an indication of the order of difference to be expected. Its calculation in this simple way makes several assumptions. The first is that water vapour permeation through plastics depends only on the vapour-pressure difference across the barrier. This may be true for gases which do not interact with the plastic, but will not in general be exactly true for water vapour and certainly not for hydrophilic films such as polyamides. Prediction of the effect of temperature follows from the basic theory that permeation is controlled by absorption or solution of the penetrant in the plastic, followed by diffusion through the barrier under a concentration gradient established between its surfaces. The work of Pecuia, for instance shows that solubility is temperature-dependent (Pecuia, 1962). Measurements by Becker and Heiss (1970) of water-vapour transmission rate for temperatures between −20°C and 50°C have been made with films of polyethylene, PVC, polyester and polyamide, alone and in combination. The marked drop in permeability which follows from the vapour pressures under freezing conditions shown in Table 5.3 is demonstrated by their sensitive apparatus.

It is of course quite inappropriate to calculate permeability per unit barrier thickness when coated or laminated materials are involved. It may also be misleading to do so even when homogeneous plastics are under study. Firstly, processing conditions affect permeability and are likely to differ in the extrusion of films of different thicknesses. Also, diffusion constants frequently depend on the concentration of the diffusing molecule in the film. Thickness can have an effect on the permeability constant of water vapour due to the concentration factor, as Scopp and Adakonis (1958) showed. Films of differing thicknesses, yet having the same concentration difference between their surfaces, will inevitably differ in concentration gradient.

Oxygen

As with water vapour, some methods of measuring gas permeability can be used only for sheet barrier materials, while others are applicable to both sheet materials and packs.

The simplest sheet methods using the pressure-differential method can be applied to gases other than oxygen. A pressure difference of one atmosphere is produced across the sheet of material and then the pressure increase on the low-pressure side measured as a function of time. Either a mercury mano-meter or an electrical pressure transducer can be used.

These methods are covered by standards such as BS 2782: Part 8: Method 821A: 1979; ISO 2556: 1974; ASTM D1434: 1975; and DIN 53380: 1969. They have the disadvantage, from a shelf-life prediction viewpoint, of relying on measurement with dry gas and, in most designs of apparatus, are unable to function at temperatures other than normal room conditions. Sensitivity with the manometric equipment has a lower limit of $1ml/m^2.day.atm$.

Permeability measurements with moist gases over a range of temperatures are possible with carrier-gas methods. In the well-known MoCon equipment (Figure 5.2), which is specific to oxygen or air, oxygen is allowed to flow over one side of a barrier sheet or over a container. This method is standardized in ASTM D3985–81. The reverse side of the sheet or the inside of the container is swept by a stream of nitrogen containing 1% hydrogen. This sweep gas stream collects the oxygen that permeates, and passes the gas mixture into a fuel cell where an electrical current proportional to oxygen concentration is generated. This fuel cell is required to be moist, hence the ability of this method to measure oxygen permeability to humidified gas streams (Delarsus, 1980; Meyer, 1957).

A variety of other detectors have been used in gas permeability equip-ments. Some use the effect of the difference in thermal conductivity between helium used as a sweep gas and other gases which permeate into it. Infrared absorption is used particularly in carbon dioxide permeability methods. The gas chromatography column with appropriate packing materials can be made to separate and quantify gases such as oxygen, nitrogen and carbon dioxide, dry or containing water vapour. The column can be fed with a portion of a sweep gas or a sample of gas mixture from a container extracted with a hypodermic syringe and injected into the column (Simril, 1950).

Thus a number of methods are available for gas permeability in which temperature and humidity can be varied and their influence studied. The influence of moisture content is most marked for the oxygen permeabilities of films containing polyamide and ethylene vinyl alcohol copolymer, and coated regenerated cellulose materials.

It has been reported that the oxygen permeability of polyamide increases by a factor of 5 as the relative humidity of the gas is raised from 0 to 100% r.h. (Anon, 1983; Brydson, 1966). A similar dependence is shown by the higher ethylene-content type E ethylene vinyl alcohol. The lower-ethylene type F may be 100 times more permeable to oxygen over the humidity range 0 to 100% (Peters, 1983; Ryder, 1984). For the other plastics barriers, polyethylene, polypropylene, the vinyl chloride derivatives and polyethylene terephthalate, measurements have shown that relative humidity changes have a much smaller effect on permeability.

The influence of temperature on the oxygen permeability of plastics can be

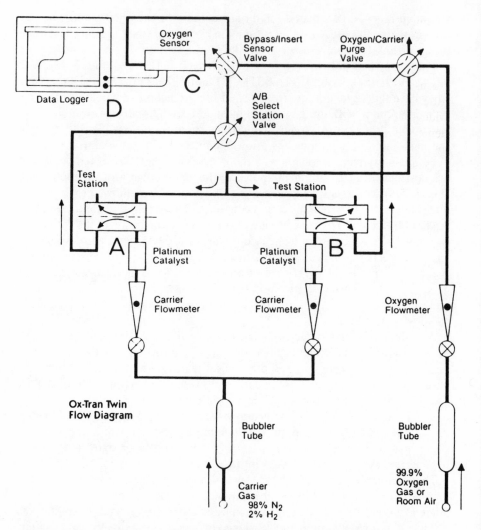

Figure 5.2 MOCON's Coulometric Oxygen Detector is a patented nickel–cadmium graphite electrode arrangement saturated with basic electrolyte to stimulate a high rate of electro-absorption of oxygen. The closed system produces a current output directly proportional to the amount of oxygen passing through the packaging material being tested which is displayed directly on the recorder. For film testing, flat $4'' \times 4''$ test specimens are clamped into special 50 cm^2 diffusion cells A and B. Both sides of the film sample are initially purged with an oxygen-free carrier gas to remove residual oxygen from the system and to desorb oxygen from the sample. When a stable zero has been established oxygen is then introduced into the upper half of the diffusion chamber. Carrier gas continues to flow through the lower half, carrying into the coulometric detector, C, any oxygen molecules which have diffused through the barrier. The strip chart recorder, D, indicates an increase in current through a special load resistor which is directly proportional to the oxygen transmission rate. After a short time, the recorder trace indicates visual vertification of the equilibrium transmission rate of oxygen through the sample barrier.

summed up by the observation that there is a linear increase in log permeability with temperature rise. An increase by a factor of 100 over the range of 0–50°C is typical of many plastics.

Given the wide range of basic plastics, the ability to produce co-extrusions, the use of laminates and coatings and the technique of metallization, it is of little consequence to tabulate values for the permeabilities of barriers. Such a wide range of permeabilities can be produced that material choice frequently depends on other considerations such as economics, appearance or packaging-line performance requirements.

Once oxygen permeability has been selected, permeabilities to carbon dioxide and nitrogen are virtually determined. Nitrogen permeability may be one-third, and carbon dioxide three or four times, the oxygen values.

Vapour permeability

The amount of information on the flavour and odour properties of packaging films has increased rapidly in recent years. The usual measurement techniques are the flow cell method with gas chromatographs to quantify the amount of volatile substance that diffuses into the sweep gas.

These measurements have shown for example that the level of permeation depends very much on the chemical affinity between the volatile substance and the plastics barrier. The most structured investigations are reported by Zobel (1982). He shows that one important feature of the permeation of volatile substances is that the rate is strongly dependent on the concentration of the permeant on the high concentration side of the barrier. The rate of permeation increases rapidly as concentration rises, a strongly non-linear dependence. Values of permeability and solubility constants thus have to be used with care in relation to the concentrations of flavours, perfumes and organic solvents in actual products.

Light transmission

The usual concern with light transmission is related to fat oxidation, a complex subject where both the type of fat and the spectral distribution in the UV and visible wavelengths is relevant to the rate of degradation (Haendler, 1978).

Most transparent plastics films transmit nearly 90% of the incident radiation down to a wavelength of 300nm. So-called 'UV absorbers' can be incorporated in some films and coatings, but other more effective ways are pigmentation, pearlization or most effectively, metallization. Printing can have a significant benefit, depending on the ink colour and density.

Shelf life prediction techniques

The essence of any shelf-life prediction technique is that it enables the period between packing and the development of an unacceptable product to be assessed in an appreciately shorter time than that period, and, it is hoped, with less overall effort than a storage trial. The major effort in a storage trial is often in the repetitive tests of product quality on samples taken at intervals from the product in store.

Full storage trials may still be necessary to provide a 'real-life' value for shelf life in confirmation of a prediction or the choice of correct combination of packaging system and product. As has been noted earlier, prediction techniques can assist selection of packaging systems or products for these storage trials.

Deterioration by loss or gain of moisture

Prediction is based on two premises:

(i) Moisture exchange is directly proportional to the total permeability of the pack; its resistance to the passage of water vapour is the reciprocal of permeability
(ii) Exchange is directly proportional to the vapour-pressure difference of water vapour across the barrier of the peak.

The permeability is usually assumed to be independent of the absolute value of the vapour pressure, and for many plastics films this is true.

The vapour-pressure difference is determined by the external vapour pressure (or relative humidity and temperature) and the internal vapour pressure, dependent on the moisture content of the contents. The various shelf-life prediction methods differ in their interpretation of this dependence of vapour pressure on the product moisture content, the absorption isotherm at the storage temperature.

Substitution method. In packaging situations where shelf life has been determined by an actual storage trial, shelf life with a new packaging material will be altered by the inverse ratio of the permeabilities of the new and old packs.

$$S(\text{new}) = S(\text{old}) \cdot P(\text{old})/P(\text{new}) \tag{1}$$

where S = shelf life and P = permeability.

This simple approach assumes that the absorption isotherm of the product is unchanged and that the storage climate, however this may vary, is constant in its effect in general terms. Another assumption is that the progress of product degradation is unaffected by the rate at which moisture absorption occurs.

Acceleration method. An increase in the storage vapour pressure can be used with caution to accelerate the rate of degradation up to the point of unacceptability of the product.

As an example, one biscuit manufacturer stores standard-size packs at 30°C and 75% r.h., recording the mass increase up to 8 days. Neglecting the first 24h of storage, the mass increase thereafter has been found to be linear. This manufacturer defines the end of shelf life as a moisture content increase of 3%, determined by organoleptic testing.

The accelerated shelf life ASL is calculated by the formula

$$ASL = (3/100 \times \text{product mass})/(\text{mass increase per day}) \qquad (2)$$

The factor 'mass increase/day' is derived from the linear portion of the 8-day storage experiment. The next stage is to multiply ASL by 4.5 to predict the unaccelerated shelf life under normally-encountered climates. In any other packaging situation, appropriate values for the acceptable rise in moisture content, in this example 3%, and the acceleration factor, 4.5, would need to be determined.

Accelerated test conditions should be used with caution, as elevated temperatures may lead to phase changes in fats and crystalline products, denaturation of proteins and changes in the absorption isotherms, for example. These possible changes mean that at the elevated temperature, the shelf life of a different product is being determined and comparison with the normal product hindered.

Linear absorption isotherm method.   Except for crystalline products such as common salt or sugar, where moisture content is low up to a critical relative humidity at which the product starts to form hydrates and/or go into solution, absorption isotherms are sigmoid in shape. As relative humidity (or water activity) rises from zero, there is an initial steep increase in product moisture content. Thereafter moisture content rises approximately linearly, finally to increase again more rapidly at humidities over about 60% r.h.

If it can be shown that the initial product moisture content (M_0) and the critical moisture content (M_c) at which the product is just acceptable fall on this approximately linear portion of the equilibrium absorption isotherm curve, then a simple calculation of shelf life is possible.

Oswin (1983) uses the analogy of the experimental rise of voltage across an electrical capacitor when charged from a constant voltage source to arrive at the equation

$$t = \frac{2.303 \cdot C}{P} \cdot \log\frac{(W_t - W_0)}{(W_e - W_0)} \qquad (3)$$

where t = shelf life in days
  C = mass of water absorbed when the exposed product is in equilibrium with the storage atmosphere, in g
  W_t = mass of product at time t, corresponding to M_C
  W_0 = mass of product at initial moisture content M_0
  W_e = mass of product at equilibrium with the storage atmosphere
  P = permeability of pack in g/day.

A very similar treatment of water exchange on the assumption of a linear equilibrium absorption isotherm was presented by Khanna and Peppas (1978). They use the normalized mass of water absorbed (n) where

$$n = m/M \qquad (4)$$

and m is the actual mass of water absorbed in a product of mass M.

The relationship between the normalized mass of water absorbed (n) and the equilibrium relative humidity (h) is assumed to be a linear equation with slope a and intercept b:

$$n = a \,.\, h + b \qquad (5)$$

Combination of equations (4) and (5) and integration between the initial and critical moisture contents gives the equation

$$t = \frac{a \,.\, M}{P} \, \ln \, \frac{(n_e - n_t)}{(n_e - n_0)} \qquad (6)$$

Substitution of values for the pack and product under study will give an estimate for the shelf life up to the critical moisture content. The factor P has to take into account the total permeability of the pack, including seals, folds and closures.

The assumption in all this prediction calculation is that the product is in equilibrium at all times with the atmosphere in the pack. That is, water can be absorbed or desorbed from the product at a rate faster than that at which vapour permeates through the barrier.

Non-linear absorption isotherm methods. Various methods have been proposed to overcome the restrictions of the assumption that the generally sigmoid absorption isotherm can be represented by a straight line between the initial and critical moisture contents. Both Oswin and Peppas have reviewed other, more complex types of equations to represent the absorption isotherm and thus facilitate shelf life prediction (Peppas and Kline, 1985).

Another approach is that of Veerraju (1970) who points out that it is possible to work with any shape of absorption isotherm without derivation of an equation to fit it. His method is graphical, and when integration of the area under a curve is required, the simple technique of 'counting the squares' or use of a planimeter is suggested.

The method starts from the rate equation:

$$dm/dt = (P/100R)(H - h) \qquad (7)$$

where the rate of change of the moisture content (m) of the product with time is related to the difference in external (H) and internal relative humidity (h), the saturation vapour pressure of water (S) and R, the resistance to moisture transmission of the pack.

Partial integration and the substitution for S/R by C, the rate of water vapour transmission for 100% r.h. differential, give

$$t = 100/C \int_{m_0}^{m} \mathrm{d}m/\mathrm{d}t \qquad (8)$$

Expressed as percentage moisture contents for a product of dry mass W, this leads to

$$\int_{m_0}^{m} \frac{100}{(H - h)} . \mathrm{d}m = t . 100C/W \qquad (9)$$

where m_0 is the initial moisture content.

The equilibrium absorption isotherm of m versus h is redrawn as $100/(H - h)$ versus m, and the area under this graph (A) derived for increasing values of m . ($A = f(m)$).

The last graph is drawn as $A = t . 100C/W + A_0$ where A_0 is the area corresponding to the moisture content, m_0 . C is determined from the mass uptake of a calcium-chloride-filled pack at the storage conditions at which shelf life is to be predicted ($A = f(t)$). A simple graphical solution of the two equations $A = f(m)$ and $A = f(t)$ for a value of m equal to the critical moisture content also gives the prediction of shelf life. This direct approach to shelf-life prediction using graphs can readily be adapted for computerization.

In a review of developments in shelf-life prediction, Oswin (1983) introduces his experience of the fitting of polynomial equations to absorption isotherms as the first stage of prediction. Giacin (1984), in a paper on moisture-sensitive drug preparations, shows that a polynomial to the power three in relative humidity will adequately fit the isotherm for his product. The equation

$$h = f(m) \qquad (10)$$

is then used to provide the successive values of h in equation (9) of Veerraju. Integration by computer, rather than squares counting, enables the predicted shelf life t to be calculated for a given initial and critical moisture content. Giacin tested the application of this technique to pharmaceutical tablets enclosed in blister packs of different materials, stored in a range of humidities. Calculated and observed moisture gains at any storage time agreed within 2–5%.

Deterioration by oxygen absorption

Many foods deteriorate due to reactions with oxygen. The situation is more complex than moisture absorption as rate of oxidation is a function of time of exposure to oxygen and also the oxygen partial pressure. The degradation is further affected by the relative humidity in the pack and exposure to light (Paul *et al.*, 1972; Anon, 1979).

Putting on one side the complications of relative humidity and light exposure by using dry, dark storage, the first need is to determine the time and partial-pressure sensitivity of the absorption of oxygen by the product. In a practical pack this oxygen may originate from gas absorbed in the product and contained in the headspace and from permeation through the pack. This permeation will occur when the internal partial pressure of oxygen is reduced either by gas flushing or by reaction of oxygen with the product.

The quantities of oxygen leading to the deterioration of oxygen sensitive foodstuffs are of the order of micrograms to milligrams per gram of product. There are two main techniques available to measure this oxygen absorption.

In the first, the foodstuff is contained in a chamber into which gas with a known partial pressure is introduced. Oxygen concentration in the headspace gas mixture is monitored by a suitable electrical oxygen analyser. The experiments would be repeated at other partial pressures. Oxygen absorption would be compared with the taste or colour change to determine the maximum acceptable absorption (Herlitze *et al.*, 1973).

The second method uses inverse gas chromatography with the column containing the foodstuff fed with the gas mixture containing oxygen at a set partial pressure. Either a pulse of oxygen containing gas is injected or a step change in oxygen partial pressure in the carrier gas made. The gas chromatography method is under development in respect of the foodstuffs to which it can be applied and the temperatures approaching ambient at which it can be used (Gilbert, 1984).

In an actual pack containing a product which absorbs oxygen, the rate at which oxygen permeates through the pack barrier (V_1) is equal to the rate of gas reacting with the product (V_2) plus the accumulation of oxygen in the free space in the pack. The value of V_1 depends on the barrier and the partial pressure difference of oxygen across it.

The volume of gas reacting with the product causes deterioration, and the first equation to produce in the shelf life prediction process is

$$M_s = f(p_0 \cdot t) \tag{11}$$

where M_s is the maximum quantity of oxygen the product can absorb and yet be acceptable for taste or colour, p_0 is the partial pressure of oxygen, and t is the time in which deterioration occurs. The average rate of oxygen absorption (V) then has to be found, also as a function of p_0, the partial pressure in the pack:

$$V = f(p_0) \tag{12}$$

The third equation that is required relates the rate of change of oxygen partial pressure in the pack to permeation and oxygen reaction with the product.

$$\mathrm{d}p/\mathrm{d}t = [DFRT(p - p_0) - RTMV_2 \cdot p_0]/V_H \tag{13}$$

where D is the oxygen permeability of package material
      F is the package surface area

R is the gas constant
T is the temperature
p is the atmospheric oxygen partial pressure
M is the mass of product
V_2 is the rate of oxygen absorption by the product
V_H is headspace volume.

The above three equations are solved by a computer program for a method of successive approximations. This programme is shown in flow chart form in Herlitze et al. (1973).

More recent developments in oxygen-dependent shelf life prediction by Khanna and Peppas apply to food water activities in the range 0.35 to 0.60. In this range, changes in water activity have little effect on the rate of oxidation. At both lower and higher water activities oxidation tends to be higher (Khanna and Peppas, 1982). Like Herlitze et al., these authors calculate the volume of oxygen which permeates into the free space in the pack due to the partial pressure difference across the barrier and the proportion of this oxygen that reacts with the foodstuff. The rate of reaction of oxygen with the food is represented by a Langmuir kinetic expression, the quantity of oxygen being proportional to the mass of product and the oxygen partial pressure.

Constant total pressure in the pack is assumed, leading to the relationship that the rate of diffusion of oxygen equals the sum of the rate of reaction and the rate of increase of oxygen in the free space. Integration of this rate equation allows the changes in internal partial pressure of oxygen to be plotted as a function of storage time. When the unacceptable rate of oxidation (related to partial pressure) is known, a shelf life can be found.

As in the work on water-vapour permeation, a permeability sorption reaction parameter is calculated (Peppas and Sekhon, 1980; Peppas and Kline, 1985). This enables the equations for rate of absorption and reaction to be solved more readily for various pack permeabilities. These authors comment that the application of this prediction technique is limited by the lack of experimental data on the kinetics of oxidation of food products. They cite the example of the deterioration of packed shrimps, monitored by colour changes, as being one product for which oxygen reaction rate parameters have been determined, and comment that shelf life where oxidation is in control is extended as the free volume in the pack is reduced.

As a general conclusion, the prediction of package shelf life for oxidation is more complex than for water-vapour permeation, and in many instances is affected by water vapour. The major difficulty is in the detailed experimental work needed to determine the oxygen absorption quantities and rates for the product to be packed.

Deterioration by combination of causes

Biscuits, which contain fat and deteriorate simultaneously by loss of crispness and development of rancidity, are frequently cited as a packaging problem needing more refined treatment than monitoring deterioration of a single quality index.

Oswin comments that it can be considered as overpacking if the deterioration by both quality indices does not reach the unacceptable limits of both at the same storage time. This may be difficult to obtain in practice, because of the available combinations of for instance water vapour and gas permeabilities in packaging materials. However, the concept makes economic sense.

The situation is more complex when the deterioration mechanisms interact, as in the loss of crispness and oxidative rancidity of potato crisps. A computer technique for shelf-life prediction, starting from the absorption isotherm for the product and oxygen permeation as set out by Becker, is possible (Quest and Karel, 1973). The technique is to produce three differential equations. The first is generally concerned with the change in oxygen partial pressure in the pack, the second with the progress of oxidation and the third with relative humidity changes during storage. These three differential equations are solved simultaneously. The interaction arises since alteration in humidity in the pack alters the rate of oxidation. The computer technique allows the oxygen partial pressure and relative humidity in the pack and the state of oxidation of the potato crisps to be plotted as a function of time.

Influence of package imperfections on shelf life

Some of the shelf-life prediction techniques described in the literature calculate a package permeability from the area of the barrier and its sheet permeability. Other techniques suggest that package permeability is determined by storage of a complete pack filled with desiccant. In this latter way the effect of folds, seals and imperfections on water-vapour transmission can be incorporated into the prediction. Continuation of this practical approach requires the effect of the variations in seal imperfection from pack to pack during a production run to be incorporated into the shelf-life prediction process, together with any other imperfections in the pack introduced during distribution and storage.

Optimization studies of heat sealing of pharmaceutical pouches in which three laminate constructions were sealed at a range of temperatures at three locations showed how leakage rate fell with selection of temperature. Significant differences between leakage rates from the same material used at three locations indicate the importance of machine adjustments (Auslander and Gilbert, 1976).

The significance of transport stresses on pouches made of seven flexible laminates has been demonstrated by both oxygen and water-vapour permeability tests, carried out before and after home trade and export schedules of drops, compression, vibration and impacts on corrugated case loads (Nielsen, 1984). In this instance, the cause of an increase in permeability was not determined. The influence of pinhole size on permeability is clear from results by Labuza (1981) from calculations using the equation

$$P = 1.77 \cdot D^2 \cdot p/L$$

where P is the permeability in mg per day through a hole of diameter Dmm in a film of thickness Lmm, with a water-vapour pressure difference of $p$mmHg. The size of pinhole in a pack can be judged by finding the pressure reduction needed to cause bubbles to emerge from a pack immersed in water in a vacuum chamber. The sensitivity of this leak detection method is comparable to the dye penetration test, but is about 1000 times less sensitive than helium leak tests (Bojkow *et al.*, 1984.)

For a given pinhole or leak size, the effect on loss of shelf life is greatest for vacuum packs and least for packs where moisture protection only is needed and diffusion under a partial pressure difference is involved.

Multiple package distribution

Whilst a simple corrugated case may in itself have little barrier effect on moisture vapour, the formation of retail cartons into collations can retard moisture ingress by the mutual covering of pack surfaces. The construction of corrugated case units into pallet loads will extend this effect to the transit and storage units. It has been shown that a carton in a collation will at worst have about twice the shelf life of an isolated pack (Paine and Turner, 1953). This effect of bulking will be lost as cases are unpacked, shelves loaded and single packs purchased. The prediction of shelf life should recognize these stages of package distribution, as otherwise overpacking can result from an under-estimation of the pack protection.

A further significant effect of palletization and bulk storage is that the mass of product and packaging in a warehouse or transit vehicle will reduce the climatic variations experienced. This is due to the thermal and moisture capacity of the stored goods.

In conclusion

Shelf-life prediction is a continuously developing subject. So far the basic mechanisms of exchange of water vapour and oxygen with the surroundings have been examined. It could be argued that migration of plasticizer and monomers from plastics containers into foodstuffs is a form of deterioration. The shelf life in this situation would be defined by the point when migration reached the legal limit for the contaminant.

Loss of a preservative gas such as carbon dioxide is another mechanism that leads to product deterioration. This can be dealt with in very much the same way as oxygen permeation, although here there is no combination with the product to take into account.

There is a growing interest in the permeation of odours and volatile substances through containers. These could be flavours or fragrances lost by the product, or unwanted odours from the surroundings permeating into the pack. The present state of the subject is that permeability data is being accumulated (Zobel, 1982; Baner *et al.*, 1984). There does not appear yet to be a direct shelf-life prediction method making use of this data. The method will be complicated by the fact that, unlike water vapour, volatiles have permeation rates depending on odour concentration as well as the odour gradient across the barrier. Related to this is the more recent work on permeation from the liquid phase as in the loss of organic solvents when stored in plastics bottles (Koszinowski and Piringer, 1986).

Whatever the circumstances in which a shelf-life prediction is being attempted, it would appear that success depends on

(i)   A thorough knowledge of the possible deterioration mechanisms of the product, including the influence of temperature, humidity and light
(ii)   A clear statement of the role played in deterioration by the various agents, e.g. water vapour
(iii)   Means of setting the acceptability limit in terms of the deterioration mechanism and these agents
(iv)   Data on the climate(s) likely to be encountered in distribution
(v)   The distribution network to be used, including the time schedule required
(vi)   Data on the permeability of possible barriers over the ranges of climate likely to be encountered and the agents associated with deterioration
(vii)   A statement as to the form and size of the pack or package to be used
(viii) Knowledge of the potential variability in the integrity of production packs, depending on the packaging material and the packaging method
(ix)   The vulnerability of the pack to distribution damage
(x)   Knowledge of the economic, health and other consequences of package failure during distribution
(xi)   A prediction method that has been tested and shown to give results that are in accord with real-life distribution and storage deterioration rates for the product in question
(xii)   An appreciation of the simplifications that have been made in the prediction method and their likely effect on the prediction result.

References

Allen, D.C. and Smith, A.T. Pira/IAPRI Int. Conf. on Packaging Technology, London, March 1972, 31.1–31.7.
Anon. (1979) *Tara* **358** (June) 337–340.
Anon. (1983) *Pkg. Japan* (Sept.) 21–25.
Apostolopoulos, D. and Gilbert S.G. (1983) *Instrumental Analysis of Foods* **2**, 51–92.
Auslander, D.E. and Gilbert, S.G.J. (1976) *Pharm. Sci.* **65** (7) (July) 1061–63.
Baner A.L. and Giacin, J.R., *Proc. Euro Food Pack*, Vienna, September 1984, 90–102.

Becker K. and Heiss, R. (1970) *Verpackungs Rdsch. Tech. Suppt.* **10**, 75–79.
Bojkow, E., Richter, C. and Potzl, G. *Proc. Euro Food Pack*, Vienna September 1984, 192–200.
Brydson, J.A., (1966) *Plastics Materials* Iliffe, London.
Cairns, J.A., Elson, C.R., Gordon, G.A. and Steiner, E.H. (1971) *BFMIRA Research Rept.* **174**.
Delassus, P.T. and Greisser, D.J. SPE Antec 1980, May 1980, 338–341.
Giacin, J.R. *4th Ann. Symp. on Pharm. Pkg.*, Rutgers University, March 1984.
Gilbert, S.G. (1984) *Adv. Chromatography* **23**, 199–227.
Haendler, H. (1978) *Rev. Choc. Conf. Bakery* **3**, 3–6, 8–10, 12.
Helen, H.J. and Gilbert, S.G. (1985) *J. Food Sci.* **50** (2) 454–458.
Herlitze, W., Heiss, R., Becker, K. and Eichner, K. (1973) *Chemic-Ingenieur Technik* **75** (8) 485–491.
Herlitze, W., Becker, K. and Heiss, R. (1973) *Verpackungs Rdsch* **24** Tech. Suppt. **7** 51–55.
Jowett, R. *Proc. Int. Conf. on Food Pkg.*, Vienna, September 1984, 331–339.
Karel, M. (1975) *Principles of Food Science Part 2: Physical principles of food preservation.* Marcel Dekker, New York, 237–263, 414.
Khanna, R. and Peppas, N.A. *Proc. Nat. Tech. Conf. Plastics Pkg.* 1978, 56.
Khanna R. and Peppas, N.A. (1982) *AIChE Symp. Ser.* **218**, 185.
Koszinowski, K. and Piringer, O. (1986) *Food Packaging and Preservation.* Elsevier, Amsterdam.
Labuza, T.P. (1982) *Food Technol.* **36** (4) (April) 92–97.
Labuza, T.P. (1981) *Cereal Foods World* **26** (7) (July) 335–343.
Lelie, H.J. (1969) *Modern Packaging* (January) 91–96.
Meyer, J.A. *et al.* (1957) *Tappi* **40**, 142–146.
Nielson, K. *Proc. Euro Food Pack*, Vienna, September 1984, 153–157.
Oswin, C.R. (1976) *Chem. and Ind.* **18** (1976) 1042–1044.
Oswin, C.R. (1983) *Package Life: Theory and Practice.* Inst. of Pkg., Leicester.
Oswin, C.R. (1983) *Food Chem.* **12** 179–188.
Paine F.A. and Paine H.Y. (1983) *A Handbook of Food Packaging.* Leonard Hill, London, 197.
Paine, F.A. (1962) *Fundamentals of Packaging.* Blackie, London, Chapter 21.
Paine, F.A. and Turner. Y.E. (1953) *PATRA Interim Rept.* **77** (June).
Paul, G., Radtka, R., Heiss, R. and Becker, K. (1972) *Fette Siefen Anstrichmittel die Ernahrungsindustrie* **74**, 336–359, 484–491.
Pecins, R.W., Ph.D. Thesis, Iowa State University, 1962.
Peppas, N.A. and Klines, D.F. (1985) *Proc. ACS Div. of Polymeric Mtls.* **52** (Spring) 579–583.
Peppas, N.A. and Khanna, R. (1980) *Polymer Sci.* **20**, 1147.
Peppas, N.A. and Sekhon, G.S. (1980) *SPE Tech Papers (ANTEC)* **26**, 681.
Peters, J.W. (1983) *Pack. Eng.* (March) 69–72.
Quest, D.G. and Karel M. (1973) *J. Food Sci.* **39** (5) 679–683.
Ryder, L.B. (1984) *Plastics Eng.* (May) 41–48.
Scopp, H.A. and Adakonis, A. (1958) *Modern Packaging* **32**, 123.
Simril, V.L. and Hersberger, A. (1950) *Modern Plastics* **27**, 95.
Sweeting, O.J. (1971) *The Science and Technology of Polymer Films*, Vol. II, Chapter 1. Wiley-Interscience, New York.
Veerraju, P.J. (1970) *Food Sci. and Tech.* **7** (Suppt. Feb.) 40–42.
Zobel, M.G.R. (1982) *Polymer Testing* **3** 133–142.

6 Cartons for liquids

ARVE J. IVERSEN

Food is packaged to preserve its quality and freshness, add appeal to consumers and to facilitate storage and distribution. (Codex Alimentarius Commission, 1985)

Introduction

Modern packaging systems for liquid foods are products from a synthesis of demands from producers, distributors, and consumers. The carton board container, or 'paper bottle', has been widely adopted because it combines hygienic protection of the product, improved working conditions in production and distribution, and consumer convenience with favourable economy.

The need for hygiene is the primary reason for retail packaging of perishable liquid food products like milk. Although this was realized more than a century ago, packaging techniques for liquid milk were slow in developing. The advent of pasteurization in the 1920s made retail packaging of liquid essential, and the returnable glass bottle was soon to become universal.

The commercial development of plastic materials, starting with polyethylene, PE, in the 1940s, opened new possibilities for improving the hygiene in liquid packaging. PE ultimately became the most frequently used thermoplastic in paper and cartonboard coating processes, also finding its use in in-plant manufacture of packages from reel stock by form–fill–seal techniques. The introduction of one-way packages for milk made the switch to self-service shops feasible, and also widened the market, as an increased number of shops were able to sell milk, thereby making it easier for consumers to buy it. In the current effort to make retailing still more efficient, the focus is on standardized packages and transport wrappings, the aim being to simplify routines and cut costs. One-way cartons are suited to meet these requirements. These developments have gradually led to a change in retail patterns in many countries, and a replacement of returnable glass bottles by single-service paper/plastic containers is observed in many countries.

Today a product distinction can be made between milk and non-milk products on the one hand, and between fresh and long-life products on the other. The main products retailed in one-way cartons are still milk and milk products, holding approximately 80% of the carton demand in liquid packaging, but a steady increase in market share for fruit juices, mineral water, sports drinks, vegetable oils and juices, soft drinks and wine is observed. This trend is likely to continue, ensuring a further potential for the paper bottle.

History and development

The carton-based packaging systems were primarily developed for milk and milk products, and there has been a constant evolution in packaging materials and overall system solutions. Other liquids requiring similar technology have benefited from and contributed to the observed evolution. The search for lower overall costs, and increasingly stringent requirements for keeping quality as a result of changes in distribution patterns and market situations, are the driving forces in this process.

A broad overview of developments starts with the paper-based container for milk, gable-top designed and patented in the USA in 1915, and the single-trip wax-coated paper cartons which were introduced commercially in 1929. Due to high overall costs, the carton package was no immediate success until Ex-Cell-O introduced their Pure Pak carton in 1936. The early design was made from semi-bleached paperboard, adhesively sealed, with a reasonable standard of hygiene and liquid-proofness being assured by immersion of the preformed carton in molten paraffin wax. The Pure Pak carton was introduced into the Scandinavian countries, produced on licence for the European market by the Norwegian Elopak Group. Although major developments have been made on the Pure Pak carton, Ex-Cell-O and Elopak retain the gable-top concept.

Meanwhile, in the early fifties, AB Tetra Pak was started off by introducing the idea of forming packages from a roll of plastic-coated paper and filling them in a continuous, closed process. The idea was considered to have a tremendous potential, and represented a radical new idea in the liquid food packaging area. With their first package, the tetrahedral Tetra Standard, polyethylene was introduced as coating, and within a few years thermoplastics totally replaced wax coatings. The Tetra Standard was introduced in 1952, and further development led to the Tetra Standard Aseptic in 1961, which was the first carton for long-life treated products.

In 1963, Tetra Pak introduced the rectangular shaped Tetra Brik to facilitate distribution. The idea of brick-shaped cartons was adopted from the German Zu-pack, but further developed to fit the Tetra Pak concept. In 1969, the Tetra Brik Aseptic was introduced, and the brick-shaped containers are now the main products from Tetra Pak. In the continous line of developments, technology has reached a point where most non-carbonated liquids may be successfully packaged in cartons, and recent developments and innovations may be characterized as adjustments and refinements of existing technology. The task now seems to be further implementation of existing technology to new markets and new product groups.

Principles of production

Filling and sealing machines for paper-based packages for liquids form two groups, those which work from a roll of packaging material, and those which work from pre-manufactured blanks, the difference reflecting the basic machine philosophy or concepts of companies like Tetra Pak and Elopak. The basic idea of Tetra Pak is that the package should be formed from a roll of packaging material, filled and sealed in a continous, closed process. The basic idea of Ex-Cell-O and Elopak is that as much as possible of the package production should be included in the converting process. Consequently, the production of blanks, being a highly specialized process, is therefore considered best performed when separated from the food packaging plant.

Converting

In the converting process, the basic paper is coated on both sides, printed, and provided with scorelines to facilitate creasing when finally forming the package. Depending on the system, the converting process may either be from reel to reel or from reel to blanks. In the latter, the converting includes punching out flat sheets of board (Figure 6.1) followed by longitudinal heat sealing, producing tubes ready for feeding to the filling machine.

Filling and sealing

When choosing a carton-based packaging system for liquids, there are three differently-shaped packages presently predominating the market, namely the gable-top, the tetrahedron, and the brick.

The filling and sealing procedure of a gable-top type package from a prefabricated blank is schematically shown in Figure 6.2. It starts with feeding of the blank from a magazine. The lay-flat tube is then unfolded and enters a mandrel where the bottom is heated with hot air. The bottom is then folded in accordance with scorelines, and pressure is applied for finishing the bottom sealing. Now an open rectangular box, it is removed from the mandrel on to a conveyer, filled with liquid and the top sealed. The top seal is performed with hot air and pressure.

The most striking feature of the tetrahedral package is the shape. The tetrahedral shape requires less packaging material than other designs, as it offers the most favorable ratio of area to volume. The forming and filling of a tetrahedral package, shown in Figure 6.3, proceeds through the following steps.

Packaging material is supplied from a reel, passes up to a bending roller and shaped into a tube with a longitudinal heat seal. Next it is filled with product, and in order to avoid foaming, the mouthpiece is located below the surface of the liquid. The transverse seals at the ends of the carton are made at right angles to each other, giving the tetrahedral shape directly. The

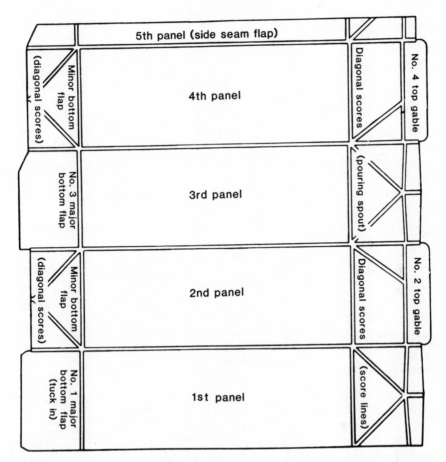

Figure 6.1 An unfolded gable-top blank is shown. Scorelines necessary for folding are indicated.

process is continuous and results in a chain of filled tetrahedral packages. As the seals are made under the liquid surface, the packages have no headspace. The transverse seals are produced by heat and pressure. First, the jaws compress the tube of material, thus excluding liquid from the sealing surface. Secondly, induction heat is supplied, melting and fusing the plastic. Pressure is maintained while cooling the seal. The outer plastic coating is then heated with a short pulse to ensure that, when opening the jaws, they do not stick to the packaging material and possibly pull the seam apart.

The production of Tetra Brik-type packages from roll-fed machines follows basically the same principles as for Tetra Standard, but the transverse seams are sealed parallel. The characteristic brick shape is formed after cutting off individual packages from the tube, by folding in the flaps and heat-sealing them.

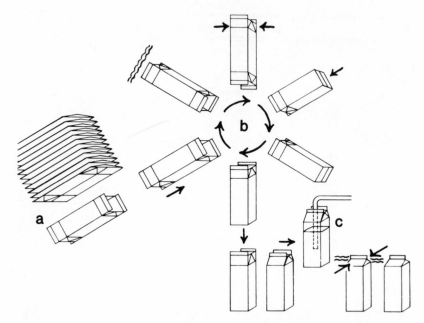

Figure 6.2 The filling and sealing procedure of a gable-top carton from prefabricated blank is shown. (*a*) The blanks are fed to the machine from a magazine, unfolded and introduced into a mandrel (*b*) where the bottom is heated, folded according to scores, and heat-sealed. Now an open box, it is removed from the mandrel, filled (*c*), and top-sealed by means of heat and pressure.

The two main types of filling machines just described have recently been supplemented by a combination of the two. The reel-fed Tetra Rex gable-top machinery has now incorporated an additional first-stage compartment in which blanks are produced. The blanks are then directly fed to a second compartment where the filling and sealing procedure follows essentially the same steps as described above.

Materials

The sandwich construction of the two common paper-based laminates used in liquid packaging are shown in Figure 6.4. If no high-gas barrier is required, the material consists of paper with a polyethylene coating on both sides. The paper layer may consist of unbleached, bleached or semi-bleached sulphate pulp, or laminates of these. The paper layer, being responsible for much of the machinability and mechanical properties of the package, requires a high' and stable quality. As the total paper consumption annually amounts to millions of tonnes, and the production process is highly specialized, only a few major producers can meet these requirements.

THE PRINCIPLE OF MAKING TETRAHEDRON PACKAGES

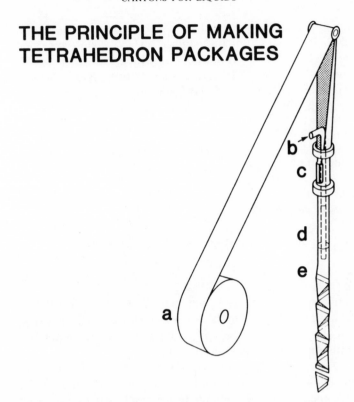

Figure 6.3 The principles of making tetahedral packages from a reel. (*a*) Packaging material on a reel is fed to the machine and formed into a tube by a longitudinal seam (*b*). The liquid enters the tube from (*b*); filling is below surface level (*d*). Finally, the package is transversely sealed (*e*) and given its final shape.

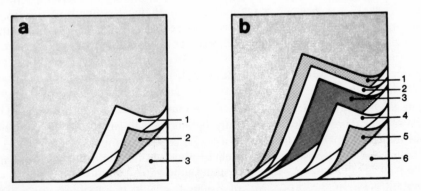

Figure 6.4 The sandwich construction of two common laminates for carton containers is shown. (*a*) Typical laminate for short-life products like fresh milk consists of (1) exterior PE, (2) paper, and (3) interior PE. (*b*) Typical laminate for long-life products consists of (1) exterior PE, (2) paper, (3) Surlyn, (4) Al-foil, (5) Surlyn, and (6) interior PE.

Additional barrier properties are usually provided by aluminium foil, laminated to the board, but the contact surface against the food remains polyethylene. The relative amount of constituents in the material may v·-y quite considerably, but a 1/1ℓ carton consists roughly of 22g paper and 6g plastics, and when Al-foil is required, approximately 1.6g of Al in addition.

Packaging materials should provide the following properties:

(i)   High hygienic standard
(ii)   Sufficient mechanical strength and internal bond
(iii)   Liquid-proofness
(iv)   Inertness to product
(v)   Light barrier
(vi)   Low gas permeability
(vii)   Sealability
(viii) Machinability in converting and filling processes.

Hygiene

The hygienic control of the packaging material must begin at the point where the paperboard is laminated in the converting process. The necessary hygienic techniques and precautions depend on type of package, materials, temperature during fabrication, handling, packing and transportation. Furthermore, the packaging material or blanks must arrive at the packaging plant with the lowest possible microbial count, and the package should be formed, filled and sealed in a manner that will eliminate any contamination.

Mechanical strength

A package should protect its contents against all normal stresses during filling, sealing, handling, and transportation. Additionally, a liquid package often acts as a dispensing container. The mechanical strength of the filled package primarily depends on the properties of the packaging material, seams and closure. The strength must be adequate for conditions of use, including internal contact with product and external exposure to ambient conditions of temperature and humidity throughout shelf life. The design of the package contributes to mechanical strength through factors such as geometrical shape and incorporation of reinforcement at points most likely to be heavily stressed. If Al-foil is incorporated in the laminate, it will contribute to total stiffness of the packaging material.

The main function of the paper layer is to impart stiffness and sturdiness to the package. The resistance of the material towards bending is a measure of the stiffness. In the processing of wood, the yield is long-fibre pulp. The quality of the fibres has the main impact on stiffness of the final sheet, other vital factors of course being paper thickness and basis weight. In the production of paper, the fibres tend to orientate in the machine direction, thus the measured stiffness will always have a higher value in the machine direction, (MD) than in the cross-machine direction, (CD). The unit of stiffness in the paper industry is traditionally that of the Taber instrument, not easily transformed into SI units. For a 1/1ℓ carton, the stiffness is about 100 Taber (CD) and 250 Taber (MD), but the ratio may vary considerably.

Future developments in paper technology may allow a more random orientation of fibres on high-speed machines, giving an improved ratio of CD to MD stiffnesses. Furthermore, there are constant efforts made in the paper industry to improve the exploitation of fibres.

Closely related to the stiffness of the carton board is the internal bond, which must be of sufficient strength to ensure that no delamination takes place. The internal bond, also termed z-direction tensile strength, is of further importance to the quality of the scorelines, i.e. the internal bonding must neither be too weak nor too strong.

The paper layer may either consist of unbleached, bleached- or semi-bleached sulphate pulp, or multiply boards consisting of laminates of these. Multiply board of lower density can have stiffness properties equal to solid board of same thickness, i.e. improved use of fibres is achieved.

Each additional step in the production of paper represents a lowering of stiffness and increased production costs. Consequently, compared to standard bleached board, the higher stiffness of the unbleached quality would allow for a favourable reduction in overall thickness of the material. Recently there was reported a total cost saving of 10.5% in favour of unbleached versus bleached in production of one-ton uncoated paper board.

It has been argued that a reduction in timber consumption may be achieved by a changeover from bleached to unbleached paper quality. However, if fibre or pulp from high-density hardwood like birch can be used in mixtures with fibre or pulp from low-density softwood like pine, a saving in timber consumption is achieved merely as a consequence of differences in density. But birch pulp requires bleaching as otherwise it is liable to cause off-taste and off-flavour problems.

Moreover, the bleached quality has a favourable white appearance, in contrast to the brown unbleached quality. Unbleached qualities are therefore often supplied with a bleached top ply or a white clay coating. Unbleached board is claimed to give products an off-taste more frequently than the bleached. This is observed mainly when the internal barrier layer merely consists of polyethylene, i.e. only a poor gas/vapour/flavour barrier to the interior.

Barrier coatings

Paper-boards have very poor water resistance and offer extremely poor barrier properties against liquids, gases and vapours. Paper-based packaging materials are therefore coated with an effective barrier to confer protection on the product as well as the paper material. In addition the coating plays an important role in the sealing of the package. Polyethylene (PE) is one of the cheapest polymers, has excellent sealing properties and exhibits high resistance to water and water vapour. It has however poor barrier properties to oxygen, carbon dioxide, aroma and flavour compounds. Since the introduction of PE, it has been unchallenged as the main coating material for cartons for fresh products like milk. PE, being easily extrudable, considered largely inert to liquid foods, and printable after treating, is regarded as an ideal plastic/food contact surface. Taking temperature and relative humidity (RH) into account, the gas permeability is given in the unit ml gas/m^2 24h atm. The rate of permeation is inversely proportional to the thickness of the material and directly proportional to area and time of exposure. Increased temperature results in increased transmission. Permeability also depends on the nature of the gas molecule, its size, shape, and polarity, but it also naturally depends on the polymer type, molecular size, crystallinity and any additives used.

The oxygen permeability of uncoated board is of the order of 800–900 ml/m^2 24h atm, and it is reduced to approximately 300 ml/m^2 24h atm by the coating process. The observed reduction is more than expected from simple resistance addition of barrier data of the paper and the PE layer separately. This is probably due to a 'plugging' of holes in the paper layer by thermoplastics achieved in the extrusion process. The effect is thoroughly described in a series of papers by Stannett and Szwarc (1962), collected in a TAPPI monograph.

When extended shelf life is desired, better barrier properties are required. Presently, this is achieved mainly by an aluminium foil, which (with no pinholes) reduces oxygen permeability to values of 0–0.5 ml/m^2 24h atm. If pinholes should by any chance exist in the Al-layer, these are also likely to be plugged by thermoplastics in the extrusion process.

The coating of thermoplastics on paperboard is performed by extrusion techniques; the plastic material is melted by heat and pressure and spread on to the board to the required thickness, by a process referred to as 'curtain coating'. High temperature allows for adhesion to the paper-board as well a proper sealing-off. Present technology allows for thickness tolerances of a thousandth of a millimetre. In contrast to PE, the ionomeric Surlyn resin has good adhesion properties to aluminium, and Surlyn and polyethylene are applied simultaneously by co-extrusion to avoid delamination.

The efficiency of thermoplastics and aluminium foil as gas barrier materials is reflected in relative material costs. Additionally, inclusion of Al-foil in the laminate adds at least one step to the converting process, and there is therefore a constant search for plastic materials with sufficiently high barrier properties at a price comparable to the Al-foil. There are today several plastic films with excellent barrier properties, but laminating them as sheets is too expensive. The future of high-barrier coating is therefore expected to be in co-extrudable polymers, featuring thin-film coating, low permeability, good heat-sealing properties, few pinholes, and good adhesion.

Seals

Cartons are sealed by heat and pressure. Heat is supplied to the sealing area to melt and fuse the thermoplastic material, usually PE. Pressure is applied while cooling the seal to ensure firm sealing. Excellent packaging materials require excellent sealing for optimum package performance. The sealing of the package must therefore be very tight. In the case of gable-top cartons, the top seal is broken when opening the package to form the pouring spout, and the top seal should therefore compromise between being tight yet easily opened. Non-tight sealing may cause deterioration due to oxidation and reinfection, whereas too-tight sealing may cause delamination when opening the package. The Pure Pak carton has a patented silicon printed area on the inner top-fin to facilitate opening. Whether this solution is better than other current methods of easy opening is here left unanswered.

Permeability to gases and water vapour

During storage, foods may deviate from their original quality in three ways: by microbiological deterioration, chemical redox reactions, or textural and chemical changes due to desiccation or hydration.

The presence of oxygen within a package often causes undesirable chemical reactions like discoloration, oxidation of unsaturated fatty acids causing rancidity problems, and loss of nutrients. Additionally, aerobic micro-organisms are often responsible for spoilage of a product. Thus it is essential to know the sensitivity of the food to oxygen, to control the oxygen content of the package, and to know the permeability of oxygen through the packaging material, seals and seams, i.e. the package as a whole. When products are packaged under an inert headspace like nitrogen or carbon dioxide, or when products evolve carbon dioxide after packaging, as in sour milk, it is essential to know the permeability features of the package to these gases. Penetration of water through the barrier layer may cause loss of mechanical strength and disintegration of the package during storage. Data on oxygen, nitrogen, carbon dioxide, and water vapour permeability should therefore be available and the requirements specified.

From a physical point of view, the permeability of packages to flavour is a phenomenon closely related to gas penetration, described above, and ruled by the same physical laws. The transfer of flavour compounds through the package may occur in either direction, from the package content to the exterior and vice versa. In the former, a reduction in characteristic aroma and flavour is observed, whereas the latter implies a penetration of volatile compounds to the product which eventually may cause off-flavours.

If the barrier layers have no micropores, the permeation of volatile substances follows from adsorption on the plastic surface, solution into the material, diffusion through the wall and finally desorption and evaporation from the opposite surface. This mechanism explains why the degree of crystallinity, the monomer content, nature and amount of additives influences permeability properties of a polymer. Selective permeability, where lipophilic polymers like polyethylene allow passage of fat-soluble compounds yet are good barriers to water and water-soluble compounds, provide one example.

If the barrier layers have many micropores or the package has a poor closure, the flow of components causing flavour transfer can proceed very quickly.

It has been realized that a pronounced gas diffusion occurs through raw cut edges of a carton package. Thus it is important, especially for long-life products, that the interior of the package has no raw edges and that the

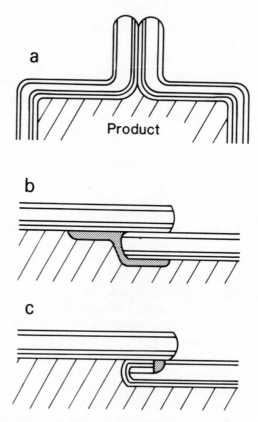

Figure 6.5 Different solutions for protection of raw edges are shown. (*a*) The common 'fin seal'; (*b*) an extra strip of plastic overlapping the internal side of the longitudinal seal as in Tetra Brik; and (*c*) in the so-called 'skiving' technique, the inner end of the carton is reduced to one-half of the original thickness, folded in and sealed to the outer end.

product is as close to being hermetically sealed as possible. Different solutions to how raw edges are protected are shown in Figure 6.5.

Gable-tops are liable to gas penetration through bottom and top seals. Bottom leakage may be minimized by the so-called Japanese bottom solution, where the possible raw edge is avoided by folding and sealing it to the exterior. The top sealing, however, is difficult to make absolutely tight, as there should be a compromise between tightness and easy opening.

Milk and many milk products have only a slight characteristic aroma. For these products, the loss or fading of flavour as a result of diffusion of volatile compounds through the package to the exterior creates no practical problem. But they also easily absorb foreign flavours, and strict requirements on permeability to volatile components, especially for products with long shelf life, are necessary.

Generally speaking, products with a neutral flavour, other conditions being identical, show faults more quickly than products with a pronounced flavour. Flavours are mostly complex mixtures of components each of which possesses a characteristic rate of diffusion with respect to the packaging material. Consequently, off-flavours sometimes occur which cannot be traced to the contaminating smell of the environment. A flavour is perceived when the concentration of the components responsible exceeds a certain threshold value. Small packages show flavour defects more quickly than large ones due to the increased ratio between the surface of the package and the product volume. Knowledge of permeability characteristics of a packaging material can thus be very useful. However, it should be stressed that such data cannot substitute for realistic performance tests of the package as a whole.

Material compatibility

Maximum inertness between packaging material and food is required. Consumer acceptance of a product depends very much on its organoleptic properties. In addition to the organoleptic changes described above, off-flavour and off-taste may be a result of migration of components originally present in the plastic/paper laminate. The direct interaction between packaging material and food involves processes like absorption, solution, and diffusion, all included in the term 'migration'.

Migration is the transfer of soluble components from the packaging material to the food, or vice versa. Hence the chemical and physical properties of the packaging material and the food, duration of contact, storage temperature, and the ratio of surface area to product volume affect the extent of migration. As no packaging material is completely inert, some migration is unavoidable. Liquid foods may be classified with respect to fat content, acidity, and alcohol content. The aqueous as well as the lipid phases of milk and milk products may act on the plastics. The ratio of hydrophilic to lipophilic components in milk and milk products varies, but the absolute fat content is less important for migration than the form and distribution in which the fat is present. Products with free fat and a continuous fatty phase are particularly active extraction media for numerous substances of low molecular weight.

A related problem is observed for aseptically-packaged fruit juices and drinks, where absorption of flavour components from the juice by the polymeric materials of the package takes place. Due to the lipophilic nature of the aromatic oil fraction and the polymeric material, the oil will diffuse into the material, causing a loss of the fresh quality.

On the other hand, food components like fats, water, alcohol, and acids may equally well diffuse into the plastics, influencing not only the migration behaviour, but also causing delamination of the packaging material.

PE, the most common contact surface between the packaging material and the liquid, is regarded as chemically stable to most food products and harmless to man.

Testing a packaging material for components affecting product flavour is done most successfully by sensory analysis. This is not only easier to carry out than chemical analysis, but also likely to give more information. Skimmed milk, apple juice, butter, and milk chocolate are used as test substances of high sensitivity for a large range of different packaging-material odours.

Finally, it should be noted that degradation due to the action of oxygen, light, and bacteria are not easily distinguished from off-flavours caused by the packaging materials.

Light

Most foods are photochemically unstable, that is, they undergo chemical changes initiated by light, often in reactions involving oxygen (Spikes, 1982). The reaction rate and products formed depend on light intensity, spectral distribution and light absorbability of the food, presence of sensitizers, temperature, and the amount of available oxygen. Photochemical effects on flavour and nutrients of liquid milk have been reviewed by Dimick (1982). The effect of light in the development of off-flavours in milk is well documented, and artificial light as well as sunlight will cause rapid deterioration. Light affects the colour and nutritive value of fruit juices and the colour of wines. Packaging materials should therefore provide the best possible light barrier. Transparent packaging materials offer only a minimum protection against photochemically-active UV and visible light. Natural unpigmented paperboard offers good protection, but light barrier may be increased by printing (Nelson, 1983), or by lamination to light-impermeable Al-foil, a solution which is required for long life products. In order to minimize light effects on fresh milk, it is recommended that the maximum permissible light transmission of the packaging material should be 8% at 500nm and 2% at 400nm (Rønkilde Poulsen, 1970). Paperboard used for liquid packaging meets this requirement with only few exceptions.

Aseptic packaging of liquid foods

In order to control and increase the microbiological shelf life of perishable liquid products like milk and fruit juices, refrigeration, heat treatment and chemical preservation may be applied. A real breakthrough for long-life treated liquids was reached in 1961, when the first Tetra Standard Aseptic packaging machines using flexible materials was introduced. Since then, a tremendous growth in aseptic packaging has been observed, making many believe that aseptics is the future of liquid packaging. When liquids are aseptically packaged, the need for keeping the chain of refrigeration unbroken is no longer essential, and shelf lives are extended to several months.

Aseptic packaging (Chapter 2) refers to a technique in which food is commercially sterilized outside the package, aseptically filled in a previously sterilized package, and sealed (Figure 6.6). Absolute sterility in an unlimited volume of product is impossible, due to the logarithmic death rate of bacteria when exposed to heat treatment. As a consequence, the concept of 'commercially sterility' is defined as a status characterized by the absence of disease-causing micro-organisms (pathogens), the absence of toxic substances, and the absence of micro-organisms capable of multiplication under normal conditions of storage and distribution.

Apart from the fact that the process should obey legal demands, it should also satisfy physical or practical demands. For the latter, any system has four main technical problems to be solved:

(i)   Commercial sterilization of the food
(ii) Commercial sterilization of the packaging material
(iii) Maintenance of asepsis of the commercially sterile food until in a hermetically-sealed container
(iv) Obtaining a hermetic seal on the package.

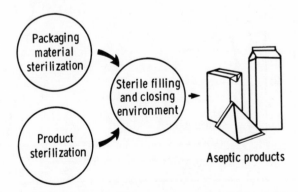

Figure 6.6 The aseptic packaging process involves separate commercial sterilization of the liquid and the packaging material, and bringing them together in a sterile environment prior to hermetically sealing.

Commercial sterilization of the liquid food is achieved by means of heat. The purposes of heat treatment are, firstly, destruction of any pathogenic organisms present in the product, making it safe for human consumption, and secondly, prolonging the keeping quality by destruction of other micro-organisms and enzymes that might cause spoilage. Every effort should be made to avoid recontamination after heat treatment. The packaging material must therefore be free from pathogens, and as free as possible from other organisms able to multiply under storage conditions.

The ultra-high-temperature (UHT) method is based on the fact that a higher temperature permits a shorter processing time. Milk is commercially sterilized with a minimum loss in organoleptic and nutritive quality if heated for 2–4 seconds at temperatures of 135–150°C, followed by rapid cooling. Other processing methods like pasteurization and sterilization expose the product to lower heat levels, but for longer periods of time (Figure 6.7), resulting in greater loss of nutritive value and an increase in 'boiled' off-taste. To ensure preservation of low microbiological activity, UHT-treated products need to be packaged aseptically.

The acceptance of hydrogen peroxide and heat as sterilizing agents by the US Food and Drugs Administration in 1981 permitted the use of aseptic packaging systems employing a wide variety of materials, such as plastics, aluminium foil, and laminates of paper to thermoplastics and aluminium foil.

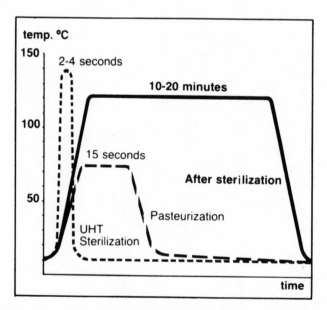

Figure 6.7 Temperature v. time comparison of three major food processing methods. Pasturization, UHT sterilization, and after sterilization (e.g. canning). The latter is not used very much in liquid packaging.

The use of hydrogen peroxide for sterilization of food-contact surfaces is covered by special regulations. They permit the use of hydrogen peroxide for food-contact surfaces of plastics, where the residual level of H_2O_2 is at or below 0.1ppm, and the sterilizing liquids are in accordance with the relevant regulations, FDA (1981) and FDA (1984).

The sterilization of packaging materials is based either on physical or chemical processes or a combination of these. Most common is the combination of hydrogen peroxide and heat, as in systems offered by Tetra Pak and PKL. Tetra Pak machines, being reel-fed, allow for complete covering of the material by aqueous hydrogen peroxide solution followed by drying off by hot sterile air in a continuous rapid process. The process is proved to be reliable, but requires a relatively high concentration of hydrogen peroxide, about 20–30%. There have been problems in sterilizing blanks in a continuous, rapid process, but these seem to be overcome in a process based on the synergistic effect of lower concentration of hydrogen peroxide, approximately 1%, in combination with UV light (Bayliss, 1979, 1982). The system patented by Liqui-Pak is presently used by Liqui-Pak and Elopak for long-life products in gable-top containers.

A perfect sterilant should be easy to apply, be part of an in-line operation, and be residue-free after application. It must have sufficient kill rate in shortest possible time, be inexpensive, and have no primary or secondary toxic effects.

Interaction between a packaging material and the food within the package is important to the shelf life of the aseptically packaged food. The ability of the packaging material to prevent oxygen diffusion into the food is the most critical factor in terms of preventing undesirable flavour changes and nutrient loss during long-term shelf life. High oxygen permeability in a fruit juice package yields poor ascorbic acid retention, increased discoloration (browning), and an overall reduction in flavour quality. Penetration of air into the package is most likely to occur through the seams and raw edges of the package, as the walls are covered with aluminium foil. Different solutions to the problem of providing seams and raw edges foil with a high gas barrier are shown in Figure 6.5.

Food processing equipment and procedures necessary to produce an

Table 6.1 Aseptic packaging systems

Systems		Shape	Sterilants
Tetra Standard Aseptic	Reel-fed	Tetrahedron	H_2O_2 + heat
Tetra Brik Aseptic	Reel-fed	Brick	H_2O_2 + heat
PKL Combibloc Aseptic	Blank-fed	Brick	H_2O_2 + heat
Pure Pak Aseptic	Blank-fed	Gable-top	H_2O_2 + heat
Liqui-Pak Aseptic	Blank-fed	Gable-top	H_2O_2 + UV
Elopak Aseptic	Blank-fed	Gable-top	H_2O_2 + UV

aseptically packaged food are changing. The number of aseptic packaging systems is increasing, as is the diversity of materials suitable for aseptic packaging. A brief survey of the major aseptic packaging systems for liquids based on paper/aluminium/plastic laminates is shown Table 6.1.

Tetra Brik Aseptic

Processing Tetra Brik Aseptic, shown in Figure 6.8, starts with the supply of packaging material to the machine from a reel. To ensure maximum gas barrier properties of the longitudinal seam and to prevent contact betwen the product and the raw paper edge, a strip of plastics is applied to one edge of the paper. After this application, the material passes through a hydrogen peroxide bath. Travelling over a forming collar and then down, the material is

TETRA BRIK ASEPTIC

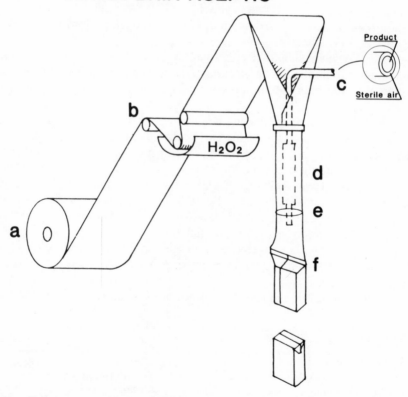

Figure 6.8 The Tetra Brik Aseptic concept is shown. (*a*) Packaging material on a reel is fed to the machine passing through a hydrogen peroxide bath (*b*), and formed into a tube by a longitudal seam. The tube heater zone is indicated by (*d*). The liquid is admitted through (*c*), and filling and sealing is performed below surface level (*e*). Finally, the package is given its final shape, (*f*).

formed into a tube by longitudinal sealing. Remaining peroxide is evaporated
in the tube heater zone, and the temperature inside reaches about 120°C,
sufficiently high to break the peroxide bond, yielding the reactive hydroxyl
radical in addition to having a heat-sterilization effect. The main sterilization
action is therefore in the tube heater zone. Product is then admitted and the
package is sealed below liquid surface level, allowing no headspace. The
sealing is performed by induction heat.

Packaging material

The aseptic packaging materials needed to provide a good package should
have all the qualities already listed, but as an additional requirement it should
be unaffected by the sterilization process. Today's packaging material is
almost entirely a paper/plastic/Al laminate, as shown in Figure 6.4b. The
aluminium foil provides satisfactory barrier properties against light and
gaseous diffusion. The laminate also offers good sealing properties when, as
in Tetra Brik, transverse seams are made by an induction heating system,
which makes sealing independent of variations in the paper layer. The
inner plastic layer usually consists of two co-extruded polyethylenes, one of
which, usually the ionomeric Surlyn, provides good adhesive properties to
aluminium. If the process includes hot-filling, Surlyn may be required on both
sides of the Al-foil layer as a precaution against delamination.

Quality control

Acceptance by consumers of long-life products is highly dependent on a
stable quality of the product. As absolute sterility is never achieved in aseptic
packaging, a maximum acceptable defective rate of 0.1% has been suggested.
Being subject to different post-process degradations, each product and
process should have clearly defined standards and specifications, without
which no optimal quality control program will work. A quality program must
cover the entire process line: raw materials, sterilizing procedures, aseptic
transfer to filler, aseptic filling operation, incubated packs, storage control,
and claim analysis. The subject of microbial quality control in aseptic
packaging has been reviewed by von Bockelmann (1982), suggested for
further reading.

Distribution

Packaging systems do not consist merely of the package itself; the package
should rather be considered as an integral part in the overall chain of
distribution, particularly for liquid foods, where billions of packages are to be
distributed annually. The package should facilitate handling at the production
plant and preliminary storage, as well as loading and unloading of the
transportation truck. It should be easily handled at the retailer by staff and
consumers. The package should in other words be optimized in all vital

aspects of ergonomics. At present, one-way cartons for liquids are distributed for retail on pallets, in crates, or in wheeled containers.

Quality demands for distribution systems for liquid packaging vary tremendously throughout the world. They depend on the market to be served, labour v. automatization costs, and the actual needs for rational handling. Retail packages for milk were traditionally heavyweight glass bottles. One-way cartons generally facilitate retail distribution, being lightweight and more easily stackable than bottles. As an example, when retailing in cartons, a truckload may consist of up to 95% liquid product and 5% packaging, whereas with bottles in crates, the truckload consists of approx. 60% liquid and 40% packaging. Returnable bottles, and other systems based on re-use, additionally increase burdens on consumers and staff by their various collection routines.

The various one-way cartons are easily dimensioned to fit the standardization requirements in transportation, such as the international pallet. A German investigation performed for Tetra Pak showed that distributing milk in $1/1\ell$ cartons on a Euro-pallet allows 720 Tetra Brik, 620 Tetra Rex Flat Top, and 470 Tetra Rex Gable-Top packages to be loaded per pallet. Less dead volume is therefore transported and costs in distribution of brick-shaped packages compared to other shapes are lower.

A few years ago, the roller (wheeled) container was introduced in the distribution of milk cartons, the intention being not a complete replacement, but rather a valuable supplement to pallets and crates. It now seems that roller containers may play an important part in the future solution of fresh milk retail deliveries. The roller container requires increased costs in the dairy, but costs are lowered in transportation and in facilitating easy handling,

Table 6.2 Types of roller container

Container type	Shelf container	Shelfless containes	
Dimensions/ volume	Gable-top (combined $1/1\ell$ and $2/1\ell$)	Flat top/ Slant top $(1/1\ell)$	Tetra Brik $(1/1\ell)$
Width	440mm	440mm	440mm
Depth	650mm	650mm	650mm
Height	1260mm	1000mm	1000mm
Packages per container	$90 - 2/1\ell$ $160 - 1/1\ell$	$160 - 1/1\ell$	$160 - 1/1\ell$
Layers	4	4	5
Packages per layer	$6 \times 4 - 2/1\ell$ $6 \times 5 - 1/1\ell$	$8 \times 5 - 1/1\ell$	$6 \times 6 - 1/1\ell$

for example in replacement in shop refrigerators and easy selection of products by consumers.

The roller container may or may not have shelves. Shelves are required for gable-top packages, and may be permanent or consist of plastic trays. Different-sized packages may be delivered on the same roller container, as packages and trays are standardized to fit modules. Table 6.2, reproduced from a survey on roller containers conducted by the Norwegian Dairy Association, shows dimensions and capacities of different roller containers.

Resources and energy

The themes of resource control and ecology considerations have attracted much attention during recent years, and are likely to be issues in the future. Presently one-way carton packages for milk, juices, and other non-carbonated beverages co-exist with one-way and returnable alternatives like the glass or plastic bottle. At first sight the returnable alternatives seem preferable from an energy as well as from an ecological point of view. But considering the entire process, from extraction of raw materials, through production and distribution to the handling of waste, this is not to be the case when compared to one-trip cartons. Investigations show that there are only minor differences, ecologically speaking, between a returnable bottle with a realistic trippage for modern retailing, and a plastic-coated paper bottle (Sundstrøm, 1982 *a*, *b*; Sundstrøm, 1985).

The constituent materials of the dominant packages for non-carbonated liquid foods, for glass bottles, cans, plastic bottles, and cartons, differ in practically every aspect: raw-material consumption, energy consumption, impact on water and air, and amount and quality of waste. The overall energy requirements for the different packages are of the same magnitude (Figure 6.9), water and air pollutants are comparable, and a larger amount of waste is generated by the returnable bottle if it does fewer than 20 trips.

Carton-based packages using wood as the basic raw material exploit a renewable resource. Present planting of trees exceeds harvesting, thus there will not be any deforestation as a result of manufacture of paper containers. Prospects for polymer as well as for aluminium supplies are good. Although oil is a limited resource, there will be no shortfall in supplies within the next century; also, only approximately 1% of total oil consumption is used for food-packaging purposes. Neither should there be any shortcoming in Al supply, as huge amounts of available raw material exist. The energy required for Al production is relatively high, making it rather expensive, but increased costs of carton/plastic/Al laminates for long-life and high-cost products may to a certain extent be balanced by a saving in distribution and refrigeration costs.

Litter and waste may be regarded as minor problems as far as PE-coated cartons are concerned. A litter problem is essentially an attitude problem, and should be treated as such. If a plastic-coated cartonboard is left in nature,

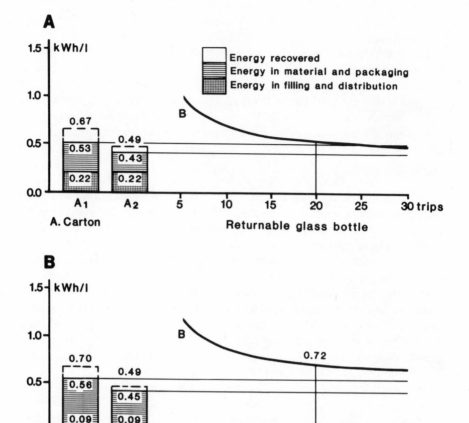

Figure 6.9 Energy counts for milk (*A*) and juice (*B*) packaged in cartons and returnable glass bottles are compared. A_1 and A_2 represent energy counts for cartons including and excluding wood energy respectively. Figures are reproduced from data given in Sundstrøm (1982 *a, b*).

complete degradation will usually require several years, the main problem being the degradation of the polymer, and biodegradable (ND) polyethylene has been discussed as an alternative outer coating. Although this solution would increase the rate of degradation considerably, it has never found application. The waste problem is the more important issue when it comes to disposabilities. Nevertheless, the waste from polyethylene-coated cartons is easily handled, does not require separate collection and may be used as a source of energy when incinerating. The combustion process of cartons yields carbon dioxide, water, and ash, and if the carton is Al-laminated, aluminium oxide forms. In Western Europe the output of waste per capita is about 250–350kg annually. If milk consumption per capita is 100ℓ a year, one-way milk cartons would be responsible for less than 1% of annual domestic waste. One tonne of milk cartons yields energy approximately equivalent to 0.4 tonnes of oil, so that incinerating them, with energy recovery, would cover the costs of collecting and disposing them.

Marketing aspects and competition

The main alternatives in liquid food packaging are glass and plastic bottles, cans, and plastic-coated cartons. A packaging system based on cartons is facilitated by low overall costs. In fact, the one-way carton container is found to be a cheaper alternative than cans and returnable glass bottles (Nunn, 1980). The profitability is largely due to low distribution costs, storage efficiency, maximum use of shelf space, and low labour costs. The main feature of cartons relative to the other systems may be summarized by the words *ease* and *economy*.

For the working environment systems based on one-way cartons are regarded as superior. Additionally, carton-based systems offer flexibility in processing different package sizes and different products. Filler systems can usually be converted to different package sizes without delay, but the same base dimension is usually required. Different products with very different processing qualities can be handled, including mineral water, nectars or fruit juices with high pulp content, wine, milk, preserved dairy products, and others.

In modern retailing the package serves the purpose of being an important informant. Being printable and having four sides available for commercials and content declaration, the carton is its own sales promoter, offering free advertisement displayed on shelves and the consumer's table. It will also serve the purpose of identifying producer and product; for example, colour codes are frequently used by dairies to distinguish various milk products.

In Western European countries, milk represented somewhat below 80% of liquid carton demand at the beginning of the 1980s, and cartons are holding about 60% of the total milk market. There is presently a decline in the total milk market, but increased consumption of cartons for milk is observed due to commercialization of long-life milk and the replacement of returnable bottles in the UK. Cartons have achieved a high penetration of the milk market in most Western countries, thus the scope for further progress may be regarded as limited outside the UK. Furthermore, fresh milk (i.e. pasteurized milk) is believed to appeal more to popular taste than sterilized milk, and markets may well turn in favour of fresh milk where proper refrigeration conditions may be achieved.

The second largest product group is for fruit juices, holding about 15% and steadily increasing, and the third largest product group is soft drinks, holding about 5%. Other products for which cartons are likely to increase in importance are wine, mineral water, vegetable oils and juices.

Selection of system

In terms of technology, the major difference between existing systems is found between reel- or blank-fed machinery. A further distinction can be made between brick-shaped and gable-top packages.

At present, paperboard containers are constructed basically of similar materials. They are based on paperboard for structural strength and graphics background, exhibit polyolefin (PE) food contact surface, and have similar barrier properties to gases and vapour. The brick-shaped container may or may not have a headspace, while the gable-top always has one, but in either style, systems allow gas flushing, giving an inert headspace and a minimum oxygen trapped within the package. They provide viable choices in sizes, price/value ratios, and utility options for most liquid foods. The brick-shaped container provides better stacking and storing properties and therefore facilitates long-life products. They are, however, limited to volumes of approximately 1/1 litre and are at present considered less 'consumer-friendly' than the gable-top which provides much easier opening, pouring, and closure properties. The gable-top cartons are often regarded as having a 'fresh-product' image.

There are many suppliers of carton-based packaging systems for liquid foods, but there are three main competitors, namely the American/Norwegian Ex-Cell-O/Elopak, the Swedish Tetra Pak, and the German PKL. Although these companies profitably sell packaging materials, their main concern has become the entire distribution system. Tetra Pak has undoubtly been the leader in this development.

Any food package has its advantages and disadvantages, and a balance must therefore be achieved to provide the protective, useful, and attractive qualities desired. Comparing developments in brick-shaped cartons and gable-tops, their respective advantages and disadvantages as packages are clearly seen. The gable-tops are considered the most consumer-friendly, with easy opening and pouring properties, but they suffer from less facility of distribution. Major efforts are therefore made to improve distribution systems, e.g. by roller containers, and the carton shape itself, e.g. slant tops, flat tops. The brick-shaped carton facilitates distribution and storage, but suffers from being less consumer-friendly with respect to opening and pouring. Efforts are made to improve these properties by different means, such as pull-tabs, application of drinking straws, etc. However, no final recommendation of system will be given here, as the selection of container and processing system is the marketer's choice based on the company's desired image, priorities and marketing strategy.

References

Bayliss, C.E. and Waites, W.M. (1979) The combined effect of hydrogen peroxide and irradiation on bacterial spores. *J. Appl. Bacteriol.* **47**, 263–269.

Bayliss, C.E. and Waites, W.M. (1982) Effect of simultanous high intensity ultraviolet irradiation and hydrogen peroxide on bacterial spores. *J. Food Technol.* **17**, 467–470.

Bockelmann, B. von (1985) Quality control of aseptically packaged food products, in *Proc. Symp. Aseptic Processing and Packaging of Food*, IUFost, Tyløsand, Sweden, 150–158.

Bockelmann, B. von (1982) *Aseptic Packaging Processing.* Tetra Pak International AB, Lund, Sweden.

Bockelmann, B. von and Bockelmann, L.I. von (1986) Aseptic packaging of liquid food products: a literature review. *J. Agric. Food Chem.* **34**, 384–392.

Dimick, P.S. (1982) Photochemical effects on flavor and nutrients of fluid milk. *Can. Inst. Food Sci. Technol. J.*, **15**, 247–256.

FDA (1981) *Fed. Reg.* **46**, 2341.

FDA (1984) *Code of Federal Regulations*, Title **21**, Part 178.1005.

FIL-IDF Bulletin (1981) New monograph on UHT milk. Document 133.

FIL-IDF Bulletin (1982) Technical guide for the packaging of milk and milkproducts. Document 143.

Nelson, K.H. and Cathcart, W.M. (1983) Analytical technique for measuring transmission of light through milk carton materials. *J. Food Protection* **46**, 309–314.

Nunn, D.W. (1980) Alternative packaging of milk — a consequence analysis. *CMI Report no. 790306-1*, Chr. Michelsens Institutt 5036, Fantoft, Norway. (Eng. summary).

Rønkilde Poulsen, P. and Blaauw, J. (1970) Influence of packaging materials on stability and organoleptic quality of milk. *IDF Bulletin*, Document 54.

Spikes, J.D. (1981) Photodegradation of foods and beverages, in Smith, K.C. (ed.), *Photochemicad Photobiological Reviews* **6**, Plenum, New York.

Stannett, V. and Szwarc, M. (1962) Permeability of plastic films and coated papers to gases and vapors, *TAPPI Monograph Series* **23**, Technical Association of the Pulp and Paper Industry, New York.

Sundstrøm, G. and Lundholm, M.P. (1982a) *Juice Packages and Energy.* G. Sundstrøm AB, S-21130 Malmø, Sweden.

Sundstrøm, G. (1982b) *Milk Packages and Energy.* G. Sundstrøm AB, S-21130 Malmø, Sweden.

Sundstrøm, G. and Lundholm, M.P. (1985) *Resource and Environmental Impact of Tetra Brik Aseptic Carton and of Refillable and Non-Refillable Glass Bottles.* G. Sundstrøm AB, S-21130 Malmø, Sweden.

7 Packaging of carbonated beverages

LOA KARJALAINEN

Historical background

Beer, besides mead, is the oldest carbonated beverage known. Brewing seems to remain more an art than an exact science, and it has not changed basically since medieval times. The history of beermaking in different countries has been described in numerous books and articles; but how the last step of the process, the packaging, has developed has created no interest among the historians. Thus, exactly when *bottling* of beer started is not known. Einbeck beer (in Worms, the town of Martin Luther) has been brewed since 1351 but bottled only from the mid 1800s; Manet showed in a painting of 1882 two bottles which apparently contained the pale ale brewed by Bass in the town of Burton-on-Trent, England; a shipment of beer from London was aboard on the steamer 'Oliva' when she sank in the Baltic in 1869, and bottles were recovered in 1974 (Jackson, 1979).

Bottling was, however, a manual or semi-manual operation until the first decade of our century, when the fully automatic manufacture of glass bottles started, or in other words, after Michael Owens had built his famous bottle-making machine. Soon after that, mechanical bottle cleaning and filling machines were invented and a historical publication of one of the breweries stated that such a machine 'had made it possible to reduce several tens of girls from the bottle washing line'.

Since the beginning of the 20th century, the bottling lines have developed with a speed that is almost beyond imagination — to wash, rinse, dry, fill, close and label more than 16 bottles a second, almost without physical labour, is a process that could not have been dreamt of 80 years ago. A range of modern containers for carbonated beverages is shown in Figure 7.1, and a modern bottling hall in Figure 7.2.

Bottled mineral waters, in contrast to beers, are only 300 years old. Boyle in 1685 published the first proposals for making carbonated mineral waters, and around 1700 champagne was invented by Dom Perignon (Moody, 1977).

It is clear that distribution of 'synthetic' (or not natural) mineral waters was made possible only when strong enough bottles and tight closures were available. The crown cork was invented in the last decade of the 19th century; before that, the closures were a problem. One solution was to use a bottle which could never stand in an upright position (Figure 7.3).

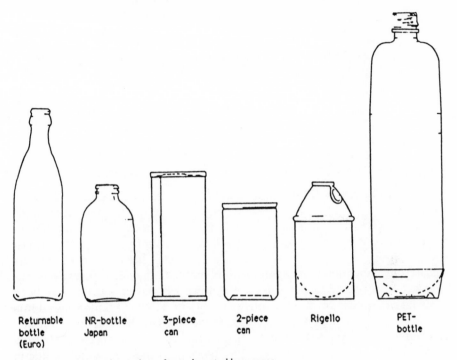

| Returnable bottle (Euro) | NR-bottle Japan | 3-piece can | 2-piece can | Rigello | PET-bottle |

Figure 7.1 A range of containers for carbonated beverages.

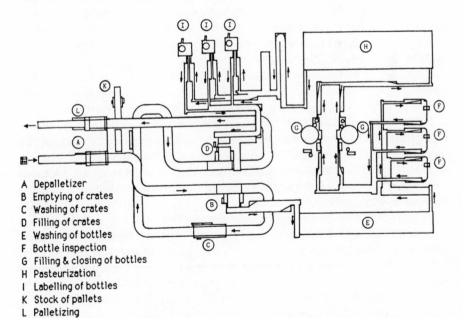

A Depalletizer
B Emptying of crates
C Washing of crates
D Filling of crates
E Washing of bottles
F Bottle inspection
G Filling & closing of bottles
H Pasteurization
I Labelling of bottles
K Stock of pallets
L Palletizing

Figure 7.2 Layout of a bottling hall.

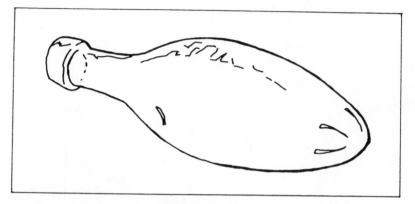

Figure 7.3 Prevention of evaporation of CO_2 from a bottle when the closure is not tight enough.

The above mentioned speed of development in packaging concerns mineral waters and flavoured carbonated beverages as well as beer. Efforts to achieve ever higher speeds on the bottling lines may be easier to understand when considering the global volume of these beverages. According to various statistics, the world consumption of beer is around 80 billion litres and that of other carbonated beverages (excluding sparkling wines) c. 88 billion litres.

From the inception of large-scale distribution of beer, the bottle has been returnable; soft drink bottles have also been returnable in many countries since their marketing began. Some governments have supported returnable systems by special taxation of non-returnable bottles or even by total prohibition of one-way bottles.

What then, has happened to the bottle itself? One could imagine that the system of returnables has contributed to large-scale standardization of bottle shapes and sizes. In fact, several standards have been created but the number of bottle shapes and sizes on the markets exceeds the number of standards manytimes. The six major franchising companies dominating the structure of the soft drinks industry (Coca-Cola, PepsiCo, Schweppes, 7-Up, Canada Dry, Dr Pepper) each have their own bottle types, varying in appearance and size (Figure 7.4). Many breweries, too, use their own bottles, although the beer bottle standards are perhaps followed better than those for soft drinks.

Not all of the present huge volume of nearly 170 billion litres of carbonated beverages is, of course, bottled in glass, whether returnable or non-returnable. Around 50 years ago the metal can came into the picture. The Gottfried Krueger Brewing Company in Newark, New Jersey, conceived the idea of filling beer in metal cans, rather than casks or bottles, for export to California. The beer quality was maintained in this way and shipping costs were low. This, together with the abolition of prohibition in 1933, resulted in a 550% increase in the company's beer sales by 1935. As a consequence, the

Figure 7.4 The well-known 20-cl bottle for Schweppes mixers is an example of a company's own bottle.

beer can was officially launched on the American market in 1935.

The new beer can was like the ordinary food can. The main difference was in a special inner varnish which protected the contents from metal flavour and which was a result of long-lasting product development by American Can Company.

To reach the contents, two holes had to be punched in the lid. For that purpose an opening tool, called the 'Church Key', was given to each customer purchasing beer cans. The can itself was of a robust construction and weighed $3\frac{1}{2}$ ounces (100g). A modern can of the same 12oz (340ml) size weighs roughly one-third as much.

Soon glass manufacturers, worried about this new, successful competitor in their traditionally exclusive markets, began to campaign against the beer can, suggesting that canned beer had a metallic flavour, that the glass bottle, because of its returnability, was more economical, and that the can required installation of new filling lines in the breweries. In reply, the can makers stated that the three big can manufacturers in the USA had overcome the problem of tainted flavour some years earlier, that canned beer in fact tasted better because it was never exposed to daylight, and that the deposit system

Figure 7.5 The cap sealed can.

for returnable glass bottles required more labour, more storage space and thus increased costs.

As one consequence of this battle, the largest brewery in the United States, Schlitz, developed a new type of can with a conical top which could be closed with a crown cork, called the 'cap sealed can' (Figure 7.5). Other can sizes were also launched — the quart (946ml) and 16oz (455ml) in 1937. The 16oz can became a success; by the same year, 37% of all American breweries used this can size, which had a 14% market share of all beer. World War II interrupted the manufacture of beer cans everywhere, but this started again by 1948, and the conical can remained on the American and European market until 1960.

In the sixties, vigorous R & D activities characterized both glass and metal beverage-packaging sectors. New opening devices were invented for tinplate cans; techniques were developed to manufacture lighter cans, which used less material for the side seam than before. The side seam was made by welding rather than soldering, as earlier. The new necked-in model allowed about

10% savings on material. Then the revolutionary two-piece can was invented, changing the whole concept of metal cans. The aluminium can was launched and began to compete so successfully with the tinplate can that within the following 15 years its market share of all beverage cans in the USA grew from 13.7% in 1970 to 94% in 1985 (Robertson, 1985).

The glass bottle industry did not remain inactive while development was taking place in the metal-can sector. As competition grew fiercer, new inventions in bottle-making and surface protection were made at an accelerating tempo. Within the past thirty years, it has been possible to considerably reduce the weight of returnable glass bottles by improving forming techniques and by developing new protective surface coating methods. The weight of a modern one-way glass bottle may be only 25% of that used in the first decades of our century. The surface of such a lightweight bottle can be coated with metal oxide and some organic compound, which together provide a good protection against scratches. As scratches on the surface of glass packages mean declining strength of the package, the development of sufficient surface protection coatings is also one key to further reduction of the glass container weight in future.

Other developments

Beside the two main materials for packaging of carbonated beverages, research and development was also carried out in other material sectors. The story of the Swedish Rigello container is a typical one of both success and failure.

Rigello was intended as a new package for carbonated beverages that could successfully compete with the traditional glass and metal packages. The following quotations, from a presentation of Rigello in 1973, reflect the way of thinking in the 60s and 70s. The whole concept of packaging design was changing from the traditional idea of 'producer — consumer' towards an integrated concept of 'producer — distribution — consumer — environment':

> Glass and metal are traditional, even historical packaging materials. They are attractive and we regard them as functional, having accepted their drawbacks as unavoidable and learnt to live with them. Product development has been tied to tradition, too, working entirely from the conditions imposed by the materials and existing production techniques...Packages for beer and soft drinks are examples of this. Glass bottles are heavy and fragile. Cans are expensive and heavy on resources of raw materials and energy...The nature of both these materials has made it necessary to adapt distribution, handling, shopping habits and environment. Most people regard this as unavoidable, even though new materials have been developed and new requirements are being made by those around us. Surely the opposite approach is the obvious one...Shopping habits and consumer preference, environmental require-ments in the widest sense, should dictate the conditions for developing materials, designs and production processes without being tied to sources of raw materials and existing production apparatus.

Rigello was made in the traditional sizes of 33, 45 and 50cl. The 33cl size had a total weight of 22.5g. It consisted of three parts: a plastic bottle, a plastic cap and a printed paper sleeve. Production could be located in the brewery, because both the bottle and the sleeve were made from material on reels with fully automatic machines. In distribution Rigello was economical because of its light weight, and environmentally it was harmless because it could easily be incinerated. A co-polymer based on acrylonitrile, called Barex 210, was used for the bottle, medium-density polyethylene for the cap and a paper/polyethylene laminate for the sleeve.

But the bottle itself caused failure in the end. The barrier properties of Barex were not adequate, and the shelf life of beer in Rigello bottles was not more than a few months, although only pasteurized beer was used. Rigello did not gain consumer acceptance. Approval was not given by the US Food and Drug Administration to use Barex for packaging of carbonated beverages (this ban was lifted in 1984, when FDA approval was given for the use of acrylonitrile co-polymers for packaging of non-alcoholic foods and beverages). The fully automatic bottle production line was too slow compared with the bottling lines in modern breweries. And, last but not least, the system was expensive. The largest brewery in Sweden, Pripps, closed down its Rigello bottling line at the beginning of the 1980s, and soon after, started cannng beer in recyclable aluminium cans.

If there is any moral in this story, it may lie in scheduling the realization of a great invention. The situation might perhaps have been more favourable for launching Rigello ten years later, when many new barrier materials had been invented, not least the modifications of polyester plastics.

The year 1977 witnessed the introduction of thermoplastic polyester as a packaging polymer for carbonated beverages. Since then, over 28 billion PET bottles (Figure 7.6) have been produced, most of them two-litre bottles for carbonated soft drinks, weighing less than 60g (Sandiford, 1985). Even beer has been introduced in PET bottles.

The astounding success of the PET bottle can be partly explained by its properties — lightweight, transparent, shatterproof, space-saving, safe. But all these consumer-appealing properties would be inadequate for the beverage producer, if the price were unsatisfactory. In large-volume bottles of today, the price as a whole is competitive, but in smaller sizes, that is, in bottles of $\frac{1}{2}$ or $\frac{1}{3}$ litre, the high energy consumption as well as the raw material costs so far make the utilization of PET uninteresting, in addition to which, the barrier properties of the ordinary PET material are not adequate for these volumes.

For bottle sizes smaller than 1 litre, another plastic material is available: polyvinylchloride (PVC). Although not yet widely used, oriented PVC has long been known and has recently been further developed. It is less expensive than PET, but its barrier properties are similar. Neither of these two plastics is able to meet the severest specifications for carbon dioxide (CO_2) retention (less than 15% loss in 120 days) applicable to highly carbonated drinks, such

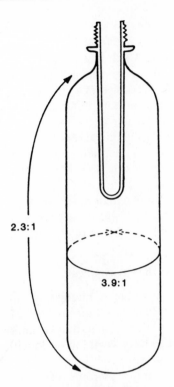

Figure 7.6 The PET bottle is made by injection stretch blow moulding where a preform, with a calibrated mouth, is blown into a mould of the final shape.

as Coca-Cola. But as not all carbonated drinks require as high a CO_2 retention, there seems to be substantial interest at present in oriented PVC bottles with capacities of 25cl and 33cl. In the United Kingdom several test marketings are in progress at the time of writing (1986).

For the packaging of highly carbonated beverages both the PET and the PVC bottle can be coated with polyvinylidenechloride (PVdC) which gives an excellent CO_2 barrier and also protects the product from oxygen intrusion. Naturally, it also raises the price, and there is still room for improvement in the quality of coating. But in the UK, PVdC-coated PET bottles have been used for beer packaging for some years already, and the critical British consumer has apparently accepted them.

PET has some disadvantages: standard PET bottles begin to distort at temperatures over 65–70°C. This means that they cannot be pasteurized by the traditional method. They can, however, be used in connection with flash pasteurization, for example, where beer is pasteurized before bottling.

Plastics have not only intruded on the territory of glass bottles; the can market now has a new competitor from plastics as well. The Petainer is a result of many years of research and development work by the Swedish company PLM Ab. Together with Metal Box, they founded the Petainer Company in the USA on a 50:50 basis, some years ago. Coca-Cola was interested in their product, which is a transparent 12-oz PET can, covered with a PVC label and with an easy-open aluminium lid. The manufacturing process, developed by PLM, consists of a unique cold-stretch-blow technique. Test marketing, carried out in three different phases in Columbus, GA., is intended to determine if Petainer in the near future will become one of the leading beverage packaging manufacturers in the world.

Another development in PET containers is a moulded PET/EVOH/PET or PET/PVdC/PET container. Nissei ASB Machinery in Japan has developed the machinery and techniques for co-injection stretch blow-moulding of bottles and containers of these materials. The high-barrier bottles prevent carbon dioxide from degassing, and these containers are expected to find a new application in the brewery industry (Ishida, 1986).

Carbonated beverages and the environment

It has been said that a package is interesting only when it contains the product. When emptied it becomes litter, and litter is harmful, polluting and expensive and it has to be disposed of as quickly and as easily as possible. The consumer does not want to have anything to do with empty packages unless he sees some advantage in it.

But litter is not the only environmental aspect of packaging. The manufacture of packages, the packaging process, distribution, and handling of solid waste are all energy-consuming phases of the cycle. The *energy consumption* as a whole in a packaging system (DsJo, 1981) might be the key question in the economics of that system. The use of raw materials (nature resources) and energy consumption are the most important environmental factors where packaging systems for carbonated beverages are concerned

Table 7.1 Total systems energy of carbonated beverage packaging, including feedstock energy, but excluding incineration. (Karjalainen, 1984, 1985)

Packaging	MWh/1 000 litres
Returnable 0.33ℓ glass bottle, bottle weight 230g, trippage (T) = 28	0.79
Non-returnable 0.33ℓ glass bottle, weight 160g, recycling rate 50%	2.70
Returnable 1.0ℓ glass bottle, weight 490g, T = 25	0.49
Three-piece 0.45ℓ tinplate can, aluminium ends, can weight 56g	5.87
Two-piece 0.45ℓ aluminium, can weight 20g	4.36
Two-piece aluminium can, as above, but with a recycling rate of 75%	2.39
1.5ℓ PET bottle, bottle weight 65g, no recycling	2.72

N.B. Because the study reflected the production and consumption conditions of 1982, both the package weights and the use of energy are somewhat too high for conditions today; e.g. the weight of a PET bottle has since decreased by about 25%.

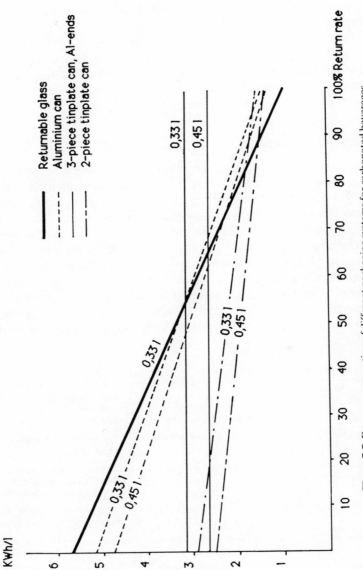

Figure 7.7 Energy consumption of different packaging systems for carbonated beverages.

(Figure 7.7). Water pollution is next to nil and air pollution is due mainly to the production of energy and to different means of transport.

In Finland, a profound study was made in 1982–85 on the environmental impact and of the economics of different packaging systems for beverages (Karjalainen, 1984, 1985). For carbonated beverages, the total energy consumption is shown in Table 7.1.

To place these figures in correct perspective, it might be mentioned that total systems energy of carbonated beverages in Finland is around 0.14% of total energy consumption of the country. Nevertheless, energy use must be taken into account when designing new packages, not only *per se*, but also because it most often influences the costs of a system.

Recycling of Beverage packages

Recycling of packages depends on consumer co-operation. Although the importance of recycling and ecology in general are now often discussed, the average consumer is still more interested in his own comfort than in ecological questions. Therefore, consumer convenience carries at least as much weight in package design as do environmental questions. For this reason it is essential that as many packaging systems as possible are designed to meet the requirements of both the consumer and of the environment. In any form of recycling of beverage packages, participation of the consumer and of the industry or society is required in order to make recycling possible.

As can be seen from Table 7.2, all package types save one require some kind of activity from the consumer. Only the tinplate can could be thrown

Table 7.2 Steps required for recycling of beverage packages

	Consumer	Society/industry
Returnable glass	Carry back to shop	Transport to bottler.
One-way glass	Take to collection bin	Transport to glass manufacturer. Crush.
Three-piece tinplate can with aluminium ends	Throw into garbage bin	Separate from dump by magnet. Remove aluminium ends and separate tin and varnish from steel
Two-piece all-tinplate can	Throw into garbage bin	Separate from dump by magnet. Separate tin and varnish from steel
Aluminium can	Take to collection bin	Compress and transport to can maker
PET bottle	Take to collection bin	Separate PVdC and PE from PET or incinerate
Petainer	Take to collection bin	Remove aluminium lid, separate PET from PVC and PVdC or incinerate

away with the usual household waste, that is, no specific steps by the consumer are needed to bring the empty package into the range of recycling process. The most economical way would, naturally, be to use all-tinplate cans which do not require aluminium separation before processing. The most troublesome package for the consumer seems to be the returnable glass bottle, because of its heavy weight compared to similarly-sized packages made of other materials, and because of the money transactions usually tied to the returning process. Depending on the location of the collection bins or centres, taking other packages to them is made more or less easy for the consumer.

The recycling value of different materials is highest for *aluminium*, where more than 90% of energy can be saved by melting and reprocessing the metal. Market resistance by environmentalists to the aluminium can was expected when it was introduced in the USA in the early 1960s. This group predicted that the 'non-recyclable' aluminium can would litter the countryside. In response to the environmental resistance, the aluminium industry initiated a recycling programme in 1970 to collect cans from consumers. Now, 16 years later, over 50% of aluminium beverage containers marketed in the USA are being recycled, and the recycling is now a great marketing benefit for this container (Dichting, 1986). In Sweden, a new recycling programme was introduced in 1980 for aluminium beverage cans. It has now been operating on full scale for two years, and the recovery rate of cans is over 60%, but the industry is optimistic about raising this rate to 70–80%. The system was planned from the beginning in an integrated way, to form a complete loop (Figure 7.8).

Glass chip or 'cullet' is widely used in bottle-making, and the proportion in the furnace can be anything between 10–90% by weight without damaging the quality of glass. Saving of energy is respectively 2.5–22.5% (Aktion Saubere Schweiz, 1984). Some glass factories even import glass chip, if it cannot be collected from their own countries.

The consumer is used to handling returnable bottles and, so far, has mostly faithfully carried the quite heavy bottles back to the shop, where he reclaimed the deposit he had paid earlier. Signs of breakdown of the system have, however, started to show, as in general the trippage (the number of trips for one bottle during its lifetime) has decreased (Table 4.3).

Collecting of non-returnable glass bottles is already well established in some countries (e.g. the Federal Republic of Germany and Switzerland, where 50–60% of glass packages are recovered) but is only starting in many others; for instance in the UK the recovery rate at present is not more than 8%. Most governmental and private organizations have realized that for glass collection to succeed, it has to be supported by efficient promotional campaigns and educational measures.

Of *plastics* materials, PET for example can be reprocessed and used for different kinds of stuffings, plates, piping, fibres, etc. Collection of plastic

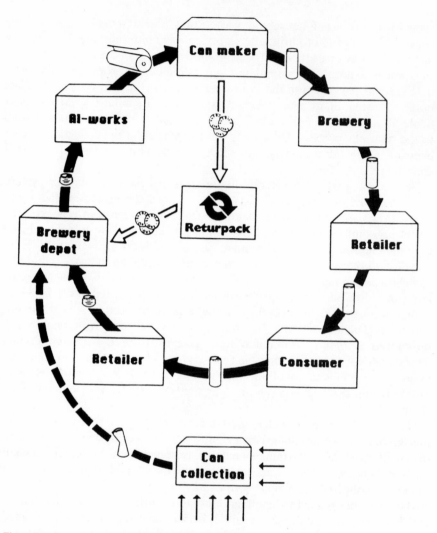

Figure 7.8 Recycling system for aluminium cans in Sweden.

waste is important because most plastic materials do not decompose, whereas for glass and metal the most important reasons for recycling are more economical than environmental. Raw material availability and costs, manufacturing cost (related to the amount of energy needed) and costs of waste handling have made it feasible to recycle metal and glass packages. They are also more interesting to the package manufacturing industry, because glass and metal packages can be recycled in a closed loop (Figure 7.8). Used PET bottles, on the other hand, cannot be used as raw material for new PET

Table 7.3 Changes in the trippage of some carbonated beverage packages

	USA (OECD, 1978)		UK (Fischer, 1983)	
	Beer	Soft drinks	Beer	Cider
1947	32	24		
1960	25	18		
1970	17	12		20
1972			18	10
1973	14.6	10	15	10
1974			13	9
1980(f)	11	7		
1985(f)	9.5	5.7		

f = forecast

bottles at present, because reprocessed PET material is unsuitable for food product packages.

Technologies exist for the separation of PET from PVC or PVdC, and such technologies are commonly used in those countries where the PET packages are collected; for instance in the USA, where PET has been marketed for nine years, recycling started six years ago. To date, about 50 000 tons of PET has been recycled. This represents a recycling rate of over 20% (Dichting, 1986). Another way of utilizing plastics waste is incineration connected with energy recovery. The BTU value of plastics is approximately twice the BTU value of mixed solid waste. As finding markets for collected plastics waste, which is mostly a mixture of different plastic types, has proved to be the problem area, research efforts have been directed towards finding uses for mixed plastic waste. Methods have now been developed to recycle this kind of waste, and the products could be pylons, sewer gratings, pipes, and signposts (Richards, 1986).

Recycling of *tinplate* requires a laborious process: varnish and tin must be separated from steel, after which both metals can be processed for re-use in metal factories. The profitability depends on the volume of collected tinplate on one hand, and on the quality of the processed steel on the other. Centralized separation and processing plants have been established in Europe, located mainly next to municipal dumps. Methods have been developed to recover high-quality steel raw material from incinerated scrap, which can be used to manufacture new cans — in other words, the closed loop of recycling is also possible for tinplate. Using all-tinplate cans, that is, two-piece cans with tinplate lids, makes the separation both easier and more economical. Collecting cans from dumps with magnets enables additional recovery of other magnetic metals as well. In the Federal Republic of Germany where 88% of the beverage cans are tinplate, recovery of steel from municipal waste is more than 2.5 times higher that the total steel in all beverage cans — a 270% recycling rate which is still increasing (Cannon, 1985).

Economics

In the USA, the virgin *raw material prices* for a 16oz (455ml) container are: glass, 1.4 cents; aluminium, 2.6 cents; PET, 3.5 cents (Dichting, 1986) (tinplate is no longer used for beverage cans in the USA). For the time being, PET seems to be by far the most expensive material. But the price of raw materials of the consumer package is naturally not the only cost factor in a beverage packaging system. In the above-mentioned study, the *total costs* of a packaging system were also calculated, as follows.

Total costs of a packaging system can be divided into

(i)   *Materials cost*, including consumer and transport packaging, waste, storage and premises
(ii) Costs of the *packaging process* including wages, energy, storage, machinery and equipment, supervising, premises, handling of returned packages
(iii) *Distribution costs* where storage, transport and return transports as well as all costs of the retailer have to be calculated
(iv) *Consumer's costs* caused by carrying, storing and returning or disposing of packages
(v)   Costs of *handling of package waste*, or outer costs.

The price indices in Table 7.4 reflect Finnish conditions, but they could be taken as indicative anywhere, especially regarding the relationship between returnable and one-way glass bottles (Karjalainen, 1985).

Table 7.4 Relative costs of different carbonated beverage packages

	Price index
Returnable 0.33ℓ glass bottle, bottle weight 230g, trippage (T) = 28	100
Non-returnable 0.33ℓ glass bottle, weight 160g, recycling rate 50%	92
Returnable 1.0ℓ glass bottle, weight 490g, $T = 25$	88
Three-piece 0.45ℓ tinplate can, aluminium ends, can weight 56g	118
Two-piece 0.45ℓ aluminium, can weight 20g	88
Two-piece aluminium can, as above, but with a recycling rate of 75%	100
1.5ℓ PET bottle, bottle weight 65g, no recycling	108

The price index shows clearly that the total costs of the 0.33ℓ non-returnable bottles are more favourable than those of returnables when a recycling rate of 50% is achieved, despite the high trippage of returnables, by international standards, in Finland. The price index does not, however, include the handling costs of solid waste. These vary remarkably depending on the type of handling but, on average in Finland, would make the one-way bottle 5–7% more expensive than the returnable one.

The impact of legislation

Problems in waste handling and the concept of recycling have caused official imposition of taxes on non-returnable packages in many countries. In some countries, there is even total prohibition, for example in Denmark, where the sale of non-returnable packages for beverages is totally prohibited. Consequently, production of metal cans, which continues only for export qualities, has decreased remarkably.

Discussion on mandatory deposits for beverage packages is under way today all over the industrialized world. Most governments today have realized that recycling of materials is the only way to diminish the amount of waste in societies. Additionally, it is assumed that by supporting returnable packaging — either directly or by deposit legislation — the problem of solid waste could be solved.

The position is difficult in some countries: in Massachusetts, USA, it has been calculated that by 1988, there will be capacity for only 65% of the solid waste produced at the present rates. To solve this problem, the government has initiated a programme worth $US 100 million of which $US 10–20 million will be dedicated to recycling (Richards, 1986). To date, nine US states have enacted a deposit law, and in these states the return rate of PET bottles, which on average for the USA is around 20%, is well over 80%.

On the other hand, opposition is hardening against deposit laws, as well as against special taxes and prohibitions. The opposition is based on the opinion that, so far, deposit laws have managed to:

(i)   Run many once-thriving voluntary recycling operations out of business
(ii) Create serious storage problems for the consumer, the retailer and the beverage supplier
(iii) Cause lost sales as consumers change their buying patterns
(iv) Cost the packaging industry many thousands of jobs (Dichting, 1986).

These opposing opinions seem, however, somewhat contradictory to the facts, such as the above-mentioned figures for recovery of PET bottles.

In Europe, non-returnable glass bottles are better accepted, especially in those countries where comprehensive glass collecting systems are established. The largest and most successful collecting systems for used packages — not only glass packaging but metal cans as well — are so far those initiated and run by the relevant industries. In most cases official bodies, municipal or governmental, participate in the systems, and in some cases, the recycling systems are authentic joint ventures between the officials, manufacturing industries and the retail trade. These voluntary (and mostly economically feasible) collecting and recycling systems have anticipated expected legislation and made it unnecessary. In many countries, however, especially where recycling of packages is not widely established, legislation concerning packages is expected to become stricter in future. Legislation concerning packaging does not necessarily only set limits to using certain package types — it may also oblige industries or trade to organize recycling programmes.

Volume of Beverage packaging systems

The population of Western Europe is 50% larger than that of the United States, but consumes less packaging (Table 7.5). General trends in packaging materials have been declining for glass and tinplate, as well as for paper and folding cartons. In Europe, consumption of aluminium has risen slightly, to a greater degree for plastics.

Table 7.5 Packaging materials consumption in Western Europe and in the USA, 1985 (Pardos, 1986).

Material, 1000 tons	Western Europe	USA
Glass	12 600	14 000
Tinplate	3 400	3 5000
Aluminium	375	1 200
Plastics	5 200	5 000
Paper	1 500	3 500
Folding cartons	2 650	3 000
Corrugated board	10 000	17 000

PET especially shows dramatic growth: European consumption for 1986 is estimated to be 110 000 tons. 80% of the European consumption of PET is used for the packaging of carbonated beverages; typically, official beer statistics of 1984 (CBMB/EBIC) do not yet recognize the PET bottle as a package in itself.

Beer-drinking habits vary remarkably between different countries. Draught beer (Figure 7.9) is highly popular in the UK (nearly 80% of the total consumed) and Federal Republic of Germany, but in the USA its share is only around 20% and is declining. Returnable bottles, in turn, represent 98% of the total beer consumption in Denmark (where non-returnable packages are prohibited), 55% in Belgium and only 7% in the United Kingdom. USA is by far the biggest user of metal cans, 51.5% in 1984, an increase of 6.2% from 1980. Figures for Western Europe show that around 40% of total consumption is draught beer, returnable bottles have about the same share and the rest is divided between non-returnable bottles and metal cans. Metal cans are increasing their share at the expense of non-returnable glass bottles. In the USA, both returnable and non-returnable glass bottles are declining, the returnables faster than the one-way bottles.

In Japan, beer production has remained reasonably stable at an annual level of 4.6 billion litres. The share of draught beer has increased from 21% to 37% during the period 1980–84. Glass bottles have decreased from 68% to

Figure 7.9 On the draught beer market there are several sizes of container–apart from the 30-litre barrel shown here, containers of up to 1000 litres are used for delivery to large distributors.

Table 7.6 Consumption of carbonated soft drinks, 1979 (Houghton, 1981)

	Total (billion litres)	Per capita (litres)
North America	31.4	130
South America	15.0	45
Western Europe	14.0	40
Eastern Europe	6.5	15
Asia	16.0	7
Australasia	1.2	68
Africa and Middle East	3.7	7

48.7%, and PET bottles, starting in 1981, have increased their share from 1 to 2% in three years (Uchida, 1986).

An estimate of the total world consumption of carbonated soft drinks is shown in Table 7.6.

The European market for soft-drinks packaging has not undergone many changes, with the exception that here, too, returnable bottles have decreased and metal cans and PET bottles increased in volume. In the UK the PET bottle seems to have taken its share from both non-returnable and returnable glass bottles, growing from nil to 17% in four years (Figure 7.10).

The volume of soft carbonated beverages sold in Japan, 2.9 billion litres, is packaged in returnable glass bottles (53.5%), non-returnable glass bottles (9.9%), PET bottles (6.4%) and metal cans (30.2%). All package types except the returnable bottles have shown a growing trend. The share of non-returnable glass bottles has increased from 0.1% to 9.9%, and that of the returnable glass bottles has declined from 67.9% to 53.5% during the period 1981–84 (Uchida, 1986).

Package sizes have become bigger in all countries, but especially in the USA. Sizes of one litre and above have become popular in glass bottles also,

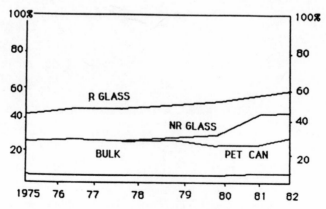

Figure 7.10 Shares of different packaging types in the soft drink industry in the UK (Can Makers' Report, 1983).

but particularly in PET bottles. A three-litre PET bottle was introduced in May 1984 in the state of Alabama. The estimated market for this size was given as 0.6 billion bottles for 1985. The estimated use of other PET bottle sizes for carbonated beverages was: half-litre and 16oz bottles, 1.3 billion; one-litre, 0.55 billion; and two-litre, 2.7 billion bottles, totalling 5.15 billion bottles (Ryder, 1984) which means approximately 8.5 billion litres of carbonated soft drinks.

Future trends or wishful thinking?

Concerning the size of beverage packages, bigger sizes than are used at present are not likely to appear in the future on a large scale. On the contrary, single serving sizes will increase, especially drinking cups. Supporters of any particular carbonated-beverage packaging system tend to predict a glorious future for their own product, but from the multitude of different trends and forecasts, some indications on future development trends seem to emerge.

Until the present, the glass bottle has been the 'unfriendliest' package regarding distribution. It is space-consuming, heavy and uses an impractical transport package which still requires much manual labour, especially when delivered to small dealers. Ideally, lightweight, space-saving beverage packages in a rolling distribution system would be substituted.

The metal can will continue its growth, more slowly than before in the USA, but faster in other countries. The competition between steel and aluminium will eventually disappear, when both metals have established their share. A natural development would be that in those countries where sources of bauxite are plentiful, such as Australia, the aluminium can will be the leader in metal cans. On the other hand, the steel can has not yet reached the limit of its technological development; the future steel can might be tin-free, two-piece, with thin walls and a steel easy-open lid. Empty cans could be collected from municipal waste dumps by means of magnets, and recycled.

The vast population of China, which seems to be committed to aluminium use, means gigantic potential for can makers and raw-material suppliers; aluminium manufacturers hopefully calculate that if the consumption of beer and soft drinks were equal to the current level in the United States, China would represent a 280-billion can market! More realistically, in Figure 7.11 the total market for carbonated beverage and beer cans is predicted to be 170 billion cans in 1995 (Robertson, 1985).

Although glass has been continuously losing its market share to metal and plastics, there will be a market for both returnable and non-returnable glass bottles in the future as well. One could speculate that the markets would be in favour of non-returnables. There is a possibility that ecological consciousness will grow enough to enable efficient recovery of non-returnable glass — and this without any deposit system. Much depends on the voluntary steps of the industry.

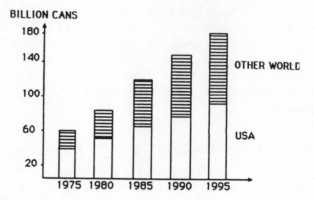

Figure 7.11 Estimated growth of metal cans for beverages.

The future of plastics still seems assured. The astounding development of the PET bottle cannot be compared with anything else in the history of packaging, yet new plastics materials, still easier to work and with still better barrier properties are certain to be developed.

As yet only imaginary, but perhaps reality in future, would be a plastic bottle or can of square shape to facilitate distribution, made of one material only to facilitate recycling, and where the material itself would allow a closed recycling loop, that is, to make plastic packages out of plastic packages.

Distribution, which today accounts for about 25% of the total systems costs, will be rationalized in different ways. The consumer package will become lighter in weight, the transport package will partly disappear and be replaced by roller pallets, and the effective use of space both in stores and in vehicles will be improved by space-saving package design.

As for collecting empty packages, in the bright future, a computerized collector will be installed at every corner. This would recognize different materials — glass, plastics, aluminium and steel — and lead them to their own channels to be crushed or compressed. All other materials not programmed would be rejected. Legislation will be needed less and less, because governments and industry together will be educating the people, organizing recycling systems and taking care of the environment. What the reality in the next decade or in the next millennium will be, however, remains to be seen for those who survive.

References

Abfall und Recycling (1984) Aktion saubere Schweiz ASS, Zurich.
Återvinning av drychesförpackningar (1980) Jordbruksdepartementet DsJo 1980:13, Stockholm.
Anon (1985) Boissons: rentabilité de l'emballage verré. *Bois* **16** (12) 18–20.
Bojkow, E. and Schlair, H. (1985) *Zur Umweltsdiskussion über Einweg-Getränkeverpackungen*, Verpackungslabor für Lebensmittel und Getränke, Wien.

Cannon, H.S. (1985) Steel processing improvements enhance steel cans, in *4th Int. Conf. on Packaging*, Michigan State University, East Lansing, Michigan.

Dichting, D. (1986) An increase in plastic recycling is needed now! in *Tenth Int. Conf. on Plastic Beverage Containers*, Ryder Associates, Inc., Whippany, NJ.

Emilson, L. (1986) The Petainer PET can technology, in *Tenth Int. Conf. on Plastic Beverage Containers*, Ryder Associates, Inc., Whippany, NJ.

Greiner, G. (1985) Geliebtes Glitzerding. *Weissblech-Reflexionen* **2**, 12–18.

Houghton, H.W. (1981) *Developments in Soft Drinks Technology 2*. Applied Science Publishers, London and New Jersey.

Ishida, O. (1986) Promising packaging products in Japan between 1985 to 1988. *Packaging Japan* **7** (31) 47–50.

Jackson, M. (1979) *The World Guide to Beer*. Quarto Publishing, London.

Karjalainen, L., Talola, M. and Viertiö, P. (1984–85) *Distribution systems of packaged beverages 1–2*. Association for Packaging Technology and Research, Espoo.

Moody, B. (1977) *Packaging in Glass*. Hutchinson Benham, London.

Oelsen, O. (1985) Der Magnet macht's. *Neue Verpackung* **7**/85, 70–72.

Pardos, F. (1986) *Packaging Industries in France and Western Europe, Present Situation and Perspectives*. Emballage, Paris.

Richards, J. (1986) Implications of recent changes in Canadian beverage packaging regulations, in *Tenth Int. Conf. on Plastic Beverage Containers*, Ryder Associates, Inc., Whippany, NJ.

Robertson, A. (1985) International beer and soft drink can market, in *4th Int. Conf. on Packaging*, Michigan State University, East Lansing, Michigan.

Ryder, L. (1984) New developments and market trends in plastic bottles, in *World Packaging Congress 1984*, Paris.

Sallenhag, J. (1985) Hanteringen lönar sig först på sikt. *Nord-Emballage* Mars, 49–51.

Sandiford, D. (1985) Bottles: what's new in two thousand years? *Plastics Today* **22**, 1–6.

Smay, G.L. (1985) Surface protective coatings — one key to reduce glass container weight, in *4th Int. Conf. on Packaging*, Michigan State University, East Lansing, Michigan.

Uchida, S. (1986) Latest trends in the PET bottle market in Japan, in *Tenth Int. Conf. on Plastic Beverage Containers*, Ryder Associates, Inc., Whippany, NJ.

8 Physical distribution today

IVANKA DIMITROVA

Background and definitions

The decade of the 1980s has been described as a 'period of flux' for the world's economies. Energy shortages, inflation, food surpluses and imbalances, together with increasing transport costs, will plague authorities in the decade ahead. These same factors will also have a significant impact on the operation of the distribution systems used by many companies because they will affect both the cost and the level of service that they can provide for their customers. Lancioni and Grashof (1981) have listed the problems to be tackled as:

(i)   The need for improved distribution management
(ii)   The need for improved distribution planning
(iii) The growth of computerized order processing systems
(iv)   A shortage of energy and critical fuels
(v)   The need for improved financial planning
(vi)   The decline in productivity
(vii) Changes in the transport regulatory structure.

Companies that have developed a well-organized system of distribution did not do this solely for cost-saving reasons. The main motive, apparently, has been to establish a systematic approach to the problem. Nowadays, when a firm becomes involved in international operations, the scope of the distribution manager's responsibilities often expands to include international distribution activities. Stock and Lambert (1982) found in a survey that the relative frequency of distribution executives with full international distribution responsibility had increased from 9.5% in 1974 to 30.5% in 1980. This trend will no doubt continue as more companies expand into global markets.

Because of its new and greater scope, the area concerned with distribution has lacked a suitable agreed name. Its substance is well identified as including transport and storage, but these functions extend both forwards and backwards. In both directions, reservoirs of goods (inventories), rates of flow, communication lags, price variations, location of facilities and market behaviour all affect optimization of customer service and distributors' profits. Several terms have been proposed to identify this area of study (Figure 8.1): materials administration, total materials systems, warehousing systems,

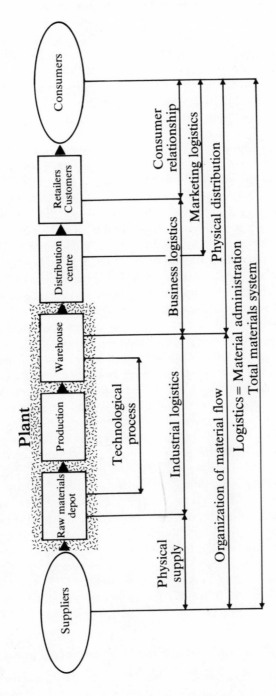

Figure 8.1 Logistics areas.

F

transportation systems, business logistics, physical distribution, and so on. Although imprecisely differentiated at first, these terms have come to be recognized and accepted through usage established by leading writers in the field. In general, 'logistics', the most inclusive of all the terms, covers the total flow of materials from origin to final destination, the systems carrying that flow, and consideration given to achieving the purpose for which they were organized. Such a total flow was divided into an incoming part, termed 'physical supply', and an outgoing part, termed 'physical distribution' (Bartels, 1982).

We can define *logistics* as 'the process of strategically managing the movement and storage of materials, parts, and finished inventory from suppliers, through the organization and on to its customers.' In Scandinavia, the term 'materials administration' has been used to denote the total concept of materials management, production control and management of the physical distribution process as an integrated whole (Ericsson, 1976, 1982; Kriens, 1976). The *production control system* is concerned with planning, scheduling and controlling the transformation of raw materials and components into finished products. The *physical supply system* is oriented towards the procurement of raw materials and components from the suppliers to the beginning of the production/assembly lines (Larsen Skjott, 1982). *Marketing logistics* is oriented towards the market and has the responsibility for moving the goods from the wholesaler's distribution warehouses to the customers and consumers. *Business logistics* is the integrated movement of material, people and related information between the source and the customer (Kriens, 1976).

According to the National Council of Physical Distribution Management (NCPDM), *physical distribution management* is the term used to describe the integration of two or more activities for the purpose of planning, implementing and controlling the efficient flow of raw materials, in-process inventory and finished goods from the point of origin to the point of consumption. These activities may include, but are not limited to, customer service, demand forecasting, distribution communications, inventory control, materials handling, order processing, parts and service support, plant and warehouse site selection, procurement, packaging, returned goods handling, salvage and scrap disposal, traffic and transportation, and warehousing or storage. Emphasis has shifted from promotion and exchange activities to the physical processes of supply, and from effectiveness to efficiency. The supplying of society with goods requires co-ordination of the physical supply and exchange functions, and the need is now greater than ever before for integration, rather than separation, of thought and effort in achieving this.

The structure of physical distribution and the importance of packaging

The descriptive structure found by Granzin (1981) using statistical processing based on a technique of factor analysis contains eight elements: system design, traffic management, transport operations, throughput operations, pallet operations, packaging, inventory control and customer relations. To delineate the structure of physical distribution we also need to know the relationships between these areas. That is to say, contemporary thinking holds physical distribution to be a system, rather than a set of independent areas of activity. Therefore, the data must be further examined to determine the relationships among the basic areas of the system. These can be represented by the correlations between the eight factors which represent the activity areas. Bivariate correlations among the resulting eight factors could then be calculated. Figure 8.2 presents an alignment of the activity areas of physical distribution based on these correlations. In two dimensions, it shows each area as a pie-shaped segment of a wheel. The correlations between adjacent areas appear as figures in the 'spokes' dividing the segments. While the packaging factor has a correlation with the other seven dimensions of physical distribution, its two highest are with inventory control (0.24) and pallet operations (0.27).

Depending on the interpretation a manager places on the word 'package', the packaging operation can represent a rather broad set of operations, all of which are concerned with bringing goods together into larger units ('make-bulk' operations). The nature of the activities involved in this area supports the broader view, as evidenced by items associated with the several different means of containing products. In addition to the first three items (package goods, open packages and design packaging), all of which clearly refer to

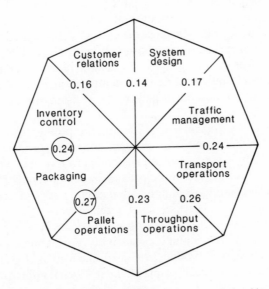

Figure 8.2 A graphic presentation of physical distribution system relationships.

packaging, the tasks in this area (containerize materials, fill bins, provide packing and dunnage to protect goods, and assign products to the proper warehouses) seem to permeate various other areas of physical distribution. Thus packaging serves as an integrative function, although one of relatively low importance, as might be expected from the sample of companies studied by Granzin which included firms other than manufacturers.

Ballou's definition (1978) of protective packaging matches the two descriptive elements of packaging and pallet operations. However, these two elements of unitizing involve a set of tasks wider than simply packaging goods. The results of Granzin's study indicate that a number of movement operations which are normally closely associated with unitizing assist in the extension of the packaging concept beyond mere concern with the package itself. The functions generally associated with packaging are: containing a product, carrying it through the distribution system, protecting it, keeping it fresh and/or usable, providing a storage (or utility) function, selling and reselling the product. Through these activities any package performs the four basic functions of containment, protection, performance and communication.

The basic reason for the existence of any package is for *containment*, so that the product can be safely moved from one place to another.

The *protection function* of the package has two aspects. The first is to protect the product from hazards — mechanical, climatic, biological — from the time it is packaged until the customer or consumer of the product has used it completely. The second aspect of protection, coequal with the first, is to protect the surrounding environment from the product in the package (Yavas and Kaynak, 1981).

The *performance function* of the package should aid in the transport, storage, selling, and use of the product. It provides the means to handle the product safely and efficiently through the channels of distribution to the point of use, and then aids in the use or consumption of the product. This function includes such things as the proper orientation of the product within the package, appropriate quantity, ease of identification, ease of removal from the package for use or ease of dispensing for consumption, ease of disposal and many more.

The *communication function* must identify the product to the prospective purchaser, inform the user of the product about the product and package features and functions, and motivate customers to buy, use, and repurchase time and again. This function of the package is of prime importance to those concerned with successful promotion and marketing of products in packages.

First of all, there are certain legal requirements. The package must, according to the laws of the state concerned, identify the product to the customer and inform them about its availability. A statement of the common or usual name of a food product and how much of it is contained in the package must be conspicuously displayed on it. Even if not required by law, it is self-evidently in the best interest of the packer to inform the customer as to what the product is, how much of it is contained in the package, and where further supplies can be obtained. Secondly, the user must be told about the important features of both the product and the package.

Thirdly, and most importantly, in a competitive market, the package must assist in selling the product. It must motivate the customer to use the product frequently. It must be designed so that it is a pleasant experience both to use the package and the product; it must look attractive in the home and in other places where it may be used. And it must create brand loyalty in the minds of the customers, so inducing them to repurchase not only that product but others from the same manufacturer.

La Londe and Czinkota (1981) studied the ranking of distribution factors in the exporting activities of US manufacturers, and their work is summarized in Table 8.1. A Likert-type scale is used (4 = very important, 1 = not important).

The table also gives the values for the standard deviations. These are quite closely clustered, which seems to indicate that firms are generally in agreement on the order of ranking the importance of physical distribution factors. It is interesting to see that executives generally believe that warehousing is of a minor importance in the context of exporting.

Figure 8.3 depicts a typical distribution organization. The distribution structure is integrated and includes the total logistics flow from the inbound to the outbound. Implicit in the arrangements is the challenge from top management that distribution management in the 1980s make a substantial profit contribution to the organization.

The integration of the distribution functions of order processing, inventory control, packaging and transportation is necessary. There is no single, simple answer or solution to the problem of organizing the logistics function in a firm. It has been claimed that a total approach to logistics can only be

Table 8.1 Ranking of distribution factors (La Londe and Czinkota, 1981)

General ranking	Factor	Importance	Std. dev.
High	Providing parts availability	3.25	0.95
	Providing technical advice	3.17	0.98
	Handling of documentation	3.07	0.95
Medium	Providing repair service	2.98	1.04
	Arranging transport	2.67	0.97
	Packaging	2.60	0.98
	Co-ordinating distribution	2.52	1.01
Low	Transport rate determination	2.32	0.97
	Obtaining insurance	2.27	0.95
	Providing warehousing	1.72	0.88

VICE PRESIDENT DISTRIBUTION				
MANAGER OF DISTRI-BUTION PLANNING	MANAGER OF DISTRI-BUTION INBOUND OPERATION	MANAGER OF CUSTOMER SERVICE	MANAGER OF DISTRIBU-TION COST ANALYSIS	MANAGER OF RETAIL DIS-TRIBUTION OPERATIONS
Forecasting	Raw materials inventory	Customer support	System cost analysis	Warehousing
Operation research	Purchasing	Customer complaints	Facility cost analysis	Materials handling
Facilities planning	Vendor selection and negotiation	Order processing		Outbound transport
Economic analysis	Inbound transport			Facilities management
Facilities analysis				

Figure 8.3 An example of a distribution organization (Lancioni and Grashof, 1981).

achieved within a matrix framework. However, application of this organizational strategy often has negative logistics consequences. In practice, there are a number of different approaches to the organization of the logistics function. First, the 'one-way' approach argues for the superiority of a given organization structure, such as having a logistics manager on the top managerial level and in a line position on a matrix organization. Alternatively, there is the 'life-cycle' approach, which identifies different stages in the logistical development of a firm and relates organizational characteristics to these stages; and finally, the 'contingency' approach, which identifies the contingencies of the organization of logistics (Persson, 1982).

New aspects of physical distribution

The problem may be approached through the application of principles and concepts; through definite, concrete computer analyses and models. The concepts and principles can provide a general shape to an economical logistics system. They aid in helping to limit the amount of in-depth analysis that must be carried out in order to implement the practice.

The important characteristics of the products being moved, from the logistical standpoint, are the volume occupied (the 'cube'), the shape and the marketing features.

The manner in which goods are distributed is most strongly influenced by how the customer perceives the product. If the product is a convenience item and customers show little brand loyalty, a rather extensive distribution system, with many stocking points and high inventory levels to provide the convenience and product availability that customers want, is desirable. Items such as foods, cigarettes and drugs are typical of this category.

On the other hand, speciality goods of the type that customers are willing to seek out and wait for when they are not immediately available, such as custom automobiles, artist's supplies and home furnishings, can be distributed in a completely different manner from convenience items. That is, little in the way of inventory needs to be maintained, and what is maintained may be highly centralized. Also, much of the cost of distribution may be charged directly to the customer. Many warehouses and private delivery services are unlikely to be a part of, nor be needed for, the distribution of such specialty products.

Other important factors are the ratio of product weight to its volume, the ratio of product value to weight, the degree to which the product can be substituted by competing products, and the degree of risk that the product could suffer in the distribution channel. A drug manufacturer or distributor is likely to distribute prescription drugs from a single stocking point for security reasons. Products such as electronic equipment, spare parts, or medical research animals, that have a high value to weight ratio, can absorb high distribution costs. Therefore, air freight can be an effective mode of transport, whereas products with low value to weight ratios (coal, iron ore, or sand) could not be moved by this high-cost mode. Similarly, products that are not well differentiated in the minds of the customer from competing products are likely to have extensive distribution networks in order to provide a high service level. Finally, products of low density, i.e. a low weight to cubic volume ratio, offer poor utilization of the storage and transport system. Such products include boats, lampshades and furniture. The strategy here is often to increase the density by shipping in the 'knock-down' state, designing special transport and storage systems, or where practical, placing production facilities close to the market place to reduce the extent of high cost shipping (Ballou, 1978).

Costs of distribution

There has been a tendency in the past to consider only the costs of transport and perhaps, of warehousing as constituting the distribution costs. More recently, however, more companies have adopted what is called the *total distribution cost* concept in respect of their distribution activities. This recognizes that many more costs are incurred through the provision of availability than those due to transport and warehousing. For example, the costs of materials handling and protective packaging should also be regarded as part of the total distribution cost, as should the costs of managing and administering the distribution system.

We can express the concept of total distribution cost in the form of an equation (Christopher, 1981):

$$TDC = TC + FC + CC + IC + HC + PC + MC$$

where TDC = total distribution costs
     TC   = transport costs
     FC   = facilities costs (depots, warehouse, etc.)
     CC   = communications costs (order processing, invoicing)
     IC   = inventory costs
     HC   = materials handling costs
     PC   = protective packaging costs
  and MC   = distribution management costs.
Figure 8.4 expresses the concept in diagrammatic form.

Various surveys have been made of the relative costs of distribution in industry, and their findings seem to suggest that, on average, distribution costs represent about 20% of sales turnover for a typical company. Clearly, averages can be misleading, and depending on the nature of the business, the actual figure in a specific instance can be much higher or lower. For example, a company in the oil business will probably find its distribution costs consideraby above the average, whilst a cash-and-carry warehouse business would have much lower total distribution costs.

Packforsk (Jonson and Balkedal, 1973) have summarized packaging costs and divided them into two main groups:

(i) costs of the packaging materials (purchase price, other costs of materials and material overheads)
(ii) Costs of handling, movement and storage (labour costs, costs of aids, costs of space).

The computer, especially the desk computer, has already made a considerable impact in many spheres of production and management, but its application in the field of physical distribution has so far been limited. For

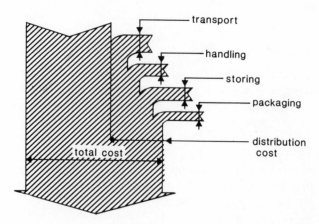

Figure 8.4 Sankay diagram of distribution costs.

example, if details of costs and available sizes of different types of packaging material, (paperboard, plastics film, shrinkwrap etc.) are considered, the financial implications of using them can be compared.

It is the interactive nature of the desk-top microcomputer, its accessibility to non-computer personnel, and the resultant ability to examine the implications of a decision in one area for other facets of the design/distribution chain, which are of importance. Some possible areas have already been highlighted, but many still remain to be investigated in a similar manner.

The essential problem facing companies in the physical distribution field is an excessively high level of costs, due mainly to a lack of co-ordination of the various elements that make up the distribution function. The first step in moving towards integrated physical distribution management is the identification of the nature and extent of these excessive costs by means of a distribution audit, and to see what scope there is for efficient distribution and handling of goods.

Figure 8.5 shows an example of an organization for effective distribution, i.e. how to reduce costs or increase profits.

The tasks of the distribution manager in the 1980s will be:

(i)   To identify areas in the company distribution system where productivity has declined
(ii) To develop solutions to improve worker output
(iii) To automate wherever possible to improve material flow and minimize duplication and excess handling.

Figure 8.6 shows the environment in which the international distribution manager must operate. The major uncontrollable elements or environments

MANAGING DIRECTOR

FINANCE	PRODUCTION	MARKETING	PHYSICAL DISTRIBUTION	MANAGEMENT SERVICES
Accounting	Line supervision	Advertising	Transport	
Budgeting	Quality control	Sales	Inventory control	
Auditing	Aggregate planning	Market research	Warehousing	
Taxation	Scheduling and despatching	Forecasting	Materials handling	
Credit approval	Inventory control	Long-range planning	Packaging	
Purchasing	Replacement & maintainance	New product development	Order processing	
	Work measurement	Customer service	Customer service	

Figure 8.5 Organization for effective physical distribution.

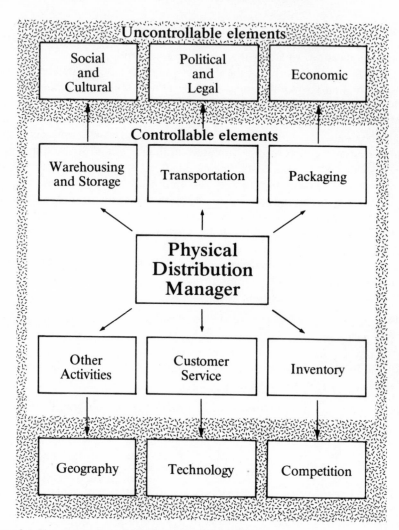

Figure 8.6 The environment of a physical distribution manager.

include the political and legal systems of foreign markets, economic conditions, degree of competition in each market, level of distribution technology available or accessible, the geography of the foreign market and the social and cultural norms of the various target markets. Any firm engaged in international distribution must still be concerned with administering the physical distribution components so as to minimize cost and provide an acceptable level of service to its customers. For example, distribution costs as a percentage of sales are much higher in Japan and the United States than in Australia or the United Kingdom (Table 8.2).

Table 8.2 Distribution costs as a percentage of sales (Gilmour and Rimmer, 1976)

Item	Country			
	USA	UK	Japan	Australia
Transportation	6.4%	5.5%	13.5%	2.5%
Receiving and despatch	1.7			1.4
Warehousing	3.7	2.5		1.8
Packaging and storage	2.6	2.0	13.0	1.7
Inventory	3.8	3.0		3.6
Order processing	1.2	1.0		2.1
Administration	2.4	2.0		1.0
Totals	21.8	16.0	26.5	14.1

Pressures from the ever-increasing complexity of technology and the increasing size of technological, commercial and social systems will provide the driving force towards the training of industrial engineers in physical distribution (Table 8.3).

Table 8.3 Areas of integrated logistics support (Ostrovsky, 1982)

Reliability and maintainability
Maintenance planning
Support and test equipment
Supply support
Transport and handling (packaging)
Technical data
Facilities
Personnel and training
Funding
Management data

There is a paucity of data routinely collected and stored on current logistic activities. It is not clear, for instance, to what extent pallets act as an alternative for other aids to movement, for unit-loads, slipsheets and etc. In this field there have been some attempts to integrate all the interrelated subsystems into an overall logistics information system (LIS) (Larsen Skjott, 1982). Some results are available in order-processing. An increasing number of companies is installing order-processing systems utilizing Electronic Point of Sale (EPOS) terminals, and others will install smart terminals, optical scanners and/or voice-operated terminals during the next decade. These will result in improved customer service, lower inventory and personnel costs and more accurate and timely order information (Mentzer, 1981).

In any event, management in inflationary conditions is not fundamentally different from that at any other time. The differences which do exist are in the pace and frequency of change and in the need to react more quickly, not only to events but also to signals in the environment, before the events occur.

Several of the following ideas and views (Farmer, 1981) are worth some consideration:

(i) Reduce stock levels generally but increase selected ranges
(ii) Reduce buying horizons on downturn
(iii) During a downturn, use resources to develop alternative materials/ suppliers in order to take advantage of competition when the upswing returns, especially where currently single-sourced
(iv) Develop a wider knowledge of the market upon which to base strategies (suppliers/competitors, potential and actual)
(v) Examine materials specifications and alternatives, being careful to use today's up-to-date data in any comparison
(vi) Examine the possibilities of minimizing cost rises through 'mutual co-operatives', e.g. between three or more parties (buyers) or on a reciprocal trading basis (buyers and sellers)
(vii) Put greater resources into the area of recycling, but with a careful appraisal of the forward scrap supply situation. If going ahead, aim to guarantee supplies with long-term contracts if there is any doubt about the quantities available
(viii) Consider booking capacity as a means of (i) ensuring supplies, (ii) helping the supplier to minimize uncertainty. (e.g. a base load in a process plant), and (iii) allowing attractive buying prices on a longer-term, possibly a fixed basis
(ix) Sound strategic buying, capacity booking or contracting during the downturn in business should result in attractive buying terms
(x) Take the opportunity during the downturn to develop databanks and associated buying information systems. Since information is a key factor in decision making, there is a need for the right data at the right time to decide on appropriate action. Speed of reaction to market trends will be a key factor in attempting to contain inflation in the procurement area
(xi) Keep up to date on counter-inflation legislation and work with suppliers to contribute to that effort — understand the services available as well as the rules
(xii) Carefully consider comparative currency values (today and into the future) when sourcing abroad — revaluation or devaluation or both can have a considerable inflationary effect
(xiii) Improve controls and cost monitoring information
(xiv) Examine intra-plant handling and transport with a view to minimizing cost.

References

Bischoff, E. and Dowsland, W.B. (1982) The desk top computer aids product design and distribution. *International Journal of Physical Distribution and Materials Management (IJPD)* **12** (1) 12–23.

Ballou, R.H. (1978) *Business Logistics*. Prentice-Hall, Englewood Cliffs, N.J.

Bartels, R. (1982) Marketing and distribution are not separate. *IJPD* **12** (3) 3–11.

Christopher, M. (1981) Logistics and the national economy. *IJPD* **11** (4) 3–29.

Ericsson, D. (1976) Transportation, warehousing and materials administration. Scandinavian Journal of Materials Administration (MA) No. 1, June.

Ericsson, D. (1982) The development of materials administration. *IJPD* **12** (3) 11–27.

Farmer, D. (1981) Insights in procurement and materials management. *IJPD* **11** (2/3) 1–134.

Gilmour, P. and Rimmer, P.J. (1976) Business logistics in the Pacific Basin. *Columbia Journal of World Business*, **11** (1) 65.

Granzin, K.L. (1981) Physical distribution from the practitioner's point of view. *IJPD* **11** (1) 40–56.

Jönson, G. and Balkedal, J. (1973) Guiding principles in the calculation of packaging costs. *Packforsk Bulletin* **26**.

Kriens, N. (1976) Applied business logistics — 'If I were a rich man'. *MA* **2** November.

Lancioni, R.A. and Grashof, J.F. (1981) Managing integrated distribution systems: The Challenge of the 80s. *IJPD* **11** (8) 3–15.

La Londe, B.J. and Cooper, M. Career patterns in distribution: profile 1980, in *Proc. 18th Annual Conference of the National Council of Physical Distribution Management*, October 13–15, 1980, 15.

La Londe, B.J. and Czinkota, M.R. (1981) The role of physical distribution in the export activity of US manufacturing firms. *IJPD* **11** (5/6) 5–12.

Larsen Skjøtt, T. (1982) Integrated information systems for materials management. *IJPD* **12** (3) 45–56.

Mentzer, J.T. (1981) Technological developments in order processing systems. *IJPD* **11** (8) 15–21.

National Council of Physical Distribution Management, Annual meeting, Chicago, Illinois, October 18–21, 1981.

Ostrofsky, B. (1981) Logistics of the third kind. *IJPD* **11** (7) 31–37.

Persson, G. (1982) Organization design strategies for business logistics. *IJPD* **12** (3) 27–37.

Stock, J.R. and Lambert, D.M. (1982) International physical distribution — a marketing response. *IJPD* **12** (2) 1–39.

Yavas, U. and Kaynak, E. (1981) Packaging: the past, present and the future of a vital marketing function. *MA* **3**, 35–53.

9 Packaging for consumer convenience

INGRID FLORY

Introduction

Practitioners in the packaging business are familiar with the most important functions of packaging: protecting the product, facilitating the handling of the product and providing information about the product and the packaging. The second aspect—facilitating the handling of the product throughout production and distribution—involves matters of great importance to the retailer and the consumer, as discussed in this chapter.

Handling of the product involves not only opening the packaging but also reclosing it where necessary, and also its storage, carriage and disposal. Surveys have shown that many consumer complaints relating to packaged foods and directed to manufacturers of consumer goods are related to the package rather than to its contents. Consumers often do not differentiate between a package and its contents, perceiving them to be an integrated entity. When a package malfunctions, the blame is transferred to the product and from there to the brand, the result being that the customer switches to a different manufacturer. The successful functioning of the package and its ergonomics are clearly not only an important part of the selling appeal of the product but also a vital area of the marketing strategy.

Non-packaged foods

A possible result of packaging malfunction is that consumers turn increasingly to products sold loose. Such trends have been seen recently in the USA and in Europe. There are examples in several market sectors, such as sweets and flour confectionery, nuts, beans and dried fruits, rice and pasta products, cured meats and delicatessen foods—all of which may be purchased either loose or in prepacked form. Other explanations for consumer preference for the non-packaged form may be environmental or health considerations, or a belief that non-packaged foods must be cheaper. Whatever the reason, there has been an increase in 'barrel and basket' displays in supermarkets and in 'fresh food' service counters.

Changing social and economic climate

The increasing demand for convenience reflects changing social and economic conditions in many countries. Most people in the Western world now have a higher disposable income and are better educated, but have less time available for meal preparation from raw foods. An increasing proportion of young adults now lack the knowledge and skill required to provide tasty, nutritious dishes from raw or unprocessed ingredients within the limits of available time and money. Convenience foods in convenient packaging can clearly satisfy a real modern need.

Another important underlying factor leading to increased support for processed food is the pressure on time experienced where all adult partners in a family unit are employed outside the home. Scarcity of time to spend in shopping and food preparation leads to an increased demand for easily-prepared packaged foods, since easily-prepared foods with a longer shelf life are now required.

The recent economic climate in the West, characterized by periods of recession and unemployment, has had profound effects on consumer behaviour, creating, amongst other things, a division between the employed, with a surplus of buying power but little free time, and the unemployed or retired, who may have adequate free time but a low discretionary income.

Attitudes towards eating itself have undergone considerable changes in the West. As a result of the spread of education and of increased media coverage of health issues, consumers are nutritionally better informed than their forebears, and may be more selective in food purchasing. Vegetarianism, for instance, appears to be on the increase in many countries. Recent controversies about the effects on health of processed foods and of 'additives' appear now, from market research, to be receding: 'natural' foods seem still to be perceived as preferable, but resistance to processed foods has diminished.

As a result of socioeconomic changes, eating habits seem to have become more flexible: fewer families now eat three formal meals per day, and more food is eaten under informal circumstances.

The food processing and packaging industries can do a great deal to satisfy the increasing need for light snack foods that are not only nutritious but also quick and easy to prepare. Products which are 'end-use-oriented' from the consumer's viewpoint offer the kind of convenience that customers require today. Such products include the 'boil-in-the-bag', 'bake-in-the-carton', 'mix-in-the-pouch' and 'serve-from-the-tray' presentations that have become increasingly popular in recent years.

Examples of these new types of convenience packaged products that can be produced by combining the newer food technologies with the latest packaging developments are examined later in the chapter.

Lifestyles and food orientation

As we have seen, the modern consumer is increasingly health-conscious, but price and convenience also play important parts in food selection. Packaging itself now becomes significant in determining purchases.

The success of many new products can be attributed to marketing, but other considerations influence consumer choice. These include techno-logical factors (developments in freezers, better cooking facilities etc.), and social factors such as changing lifestyles bringing more leisure time, more double-income family units, more eating out (frequently in fast-food restaurants); smaller families and the growth of single-person households; increased incomes; and better education. What does the consumer require as a result of these social and economic changes? Demand in the food area seems still to be for simplified preparation and convenience (Figure 9.1). In terms of packaged product manifestations, this involves frozen meals, add-hot-water soups and snacks, aseptically packaged juices, microwavable dishes, 'gourmet' foods, ice-creams in artistic variations and many new types of yoghurt and other processed dairy products. Clearly wide opportunities exist for innovative packaging and creative technology.

Figure 9.1   Example illustrating new trends for convenience foods: frozen dinners, of high quality and low calorie content, easy to prepare in microwave or conventional stove. A gourmet dish from Findus, produced in Canada and sold in Sweden.

Recent developments

Good examples of convenient packaging and product/package combinations which have long been established in the marketplace—and seem likely to continue—are squeezable plastic bottles for ketchup, plastic/paper tubs which also perform as table dispensers for butter and margarine (Figure 9.2), lighter-weight paper-based containers for milk and milk products and juices, and lighter-weight containers for carbonated beverages (made in aluminium, thinner, more uniform glass or coated plastics). The general tendency is to reduce package weight without reducing efficiency, partly for ease of handling but also as a result of increasing raw materials costs.

Figure 9.2 The Tritello tub for margarine in paper/plastic, convenient to use and store compared to the traditional foil. The container has made it possible to spread butter/ margarine right from the refrigerator. (By courtesy of Akerlund and Rausing, Lund, Sweden.)

New plastic materials

Blow-moulded packagings range from small 4 ml bottles to 30 l containers. Polyethylene is the main material, but PVC and polyester are also widely used. Plastics packaging is being increasingly used as developments in improved sealing properties become commercial.

Dairies have traditionally been large consumers of thermoformed tubs and cups. Not only can they be printed in up to five colours, but they are pleasant to eat from and can be used afterwards for seed propagation, etc., before disposal.

New multi-layer co-extruded or laminated materials based on plastics have produced possibilities for interesting food product developments incorporating more convenient packaging. Tomato ketchup, for example, has been presented in plastic containers for some years, but the higher-quality brands, without additives and preservatives, have until recently required the barrier protection given by glass. New multilayer plastics materials with greatly improved barrier characteristics now provide a realistic alternative to the glass bottle, thereby giving a lighter and more convenient package (Figure 9.3).

Figure 9.3 Tomato ketchup of high quality in a new co-extruded plastic bottle with 'safety seal', almost as tight as a glass bottle.

The tremendous growth of large ($1\frac{1}{2}$ and 2-litre) PET bottles (Figure 9.4) for soft drinks and mineral spring waters is a further example of the success of these materials in providing innovative convenience for consumers. A filled $1\frac{1}{2}$-litre bottle in PET, for example, has about the same weight as a filled 1-litre glass container and is also easier to handle and dispose of.

Figure 9.4 Rapidly growing markets—cola, and mineral and spring water in large PET bottles, light and easy to carry and handle, and quickly accepted for consumer convenience.

Nutritional supplement in Tetra Brik Aseptic

The Swedish food manufacturer SEMPER, a subsidiary company of the milk producers' cooperatives, has produced, for a number of years, a liquid nutritional supplement for patients requiring post-operative treatment, or others in need of additional nourishment. Until recently this product was distributed in the traditional metal can, but can now be packaged in the new Tetra Brik Aseptic carton (Figure 9.5). A drinking straw is attached to the carton to facilitate consumption. This new package improves both distribution and use of the product, possibly making it available to new groups of patients. The package makes the product easy to serve, consume, distribute and dispose of, since it eliminates the need for extra serving units and cleaning of dishes. This is a good example of the way packaging can combine consumer convenience with a specialized feeding need.

School milk programmes

Many countries all over the world have extensive school milk

Figure 9.5 Liquid nutritional supplement in four flavours packaged in Tetra Brik Aseptic, convenient to consume for post-operative patients.

programmes, through which children are encouraged to drink milk, and parents are made aware of the importance of the children's nutritional requirements. National programmes of this type are especially valuable in developing countries like Kenya and Singapore, but also include Japan and many European countries (Figure 9.6).

(a)

(b)

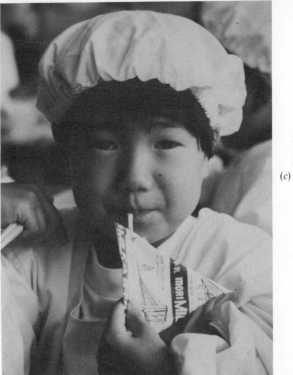

(c)

Figure 9.6   (a) Kenyan school children collecting school milk in tetrahedral Tetra Paks. (b) The Tetra Pak standard 200 ml and the Tetra Brik Aseptic which is replacing it in Kenya. (c) Tetra Pak in Japan.

Such programmes represent a major economic commitment by national or local governments, and practical distribution problems are clearly considerable. Here the packaging of individual servings of milk in cartons has made distribution cheap and convenient. In three countries in Europe, for example, the children themselves assist in the distribution.

In Denmark, school milk, plain white or chocolate-flavoured, is served every day. It is packaged in 250 ml Tetra Brik or Tetra Rex cartons kept under refrigeration in the school. Two children from each class collect the appropriate number of cartons from the store and distribute them in the classroom. The empty cartons are then deposited in large plastic sacks which go straight to the refuse collection.

A similar method of involving the children is practised in Norway, where school milk in disposable cartons was introduced in the early 1960s. The children receive their school milk five times a week and it is normally distributed to the classrooms by members of the pupil's council. Here, it is packaged in 250, 300 or 333 ml Tetra Pak cartons.

In the Netherlands, school milk was first introduced in 1935, and it has been packaged and distributed over the years in various ways. Today, the children receive their milk in 250 ml Tetra Brik cartons with a drinking straw during the morning break. Delivery to the schools is made only twice each week and it is stored in the school's refrigerated cabinets. Again the children themselves are responsible for all milk handling and storage operations in the school.

Opening and reclosing functions

In recent years there has been a great demand for packaging materials with properties that make packages easy to open and reclose. A number of solutions to these problems have been devised which are applicable to flexible packagings. Examples and their applications are shown in Figures 9.7–9.10. These materials are based on the plastic laminate Transofoil E. The convenience to the consumer lies in the ability to quickly open, yet securely reclose a flexible packaging without cutting, repackaging, pulling or tearing, thus maintaining the benefits of the packaging for the product.

Let Pak

The Let Pak food container was developed to replace the traditional tin can for heat-sterilized foods. It is made from co-extruded plastic material, and can be opened without a can opener and heated in a microwave oven. Compared with the metal can, Let Pak is lighter in weight, and its white colour is said to give it an appealing fresh and hygienic appearance (Figure 9.11).

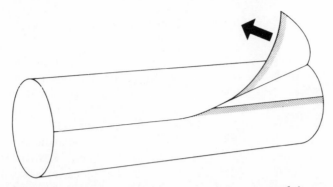

Peel-n-seel — Easy to open and just as easy to reseal.

A special coldseal application that makes the package as easy to reseal as it is to open. Peel-n-seel can be applied at the same time as the package is being printed. Suitable for all types of flexible packages printed in rotogravure.
Open and reseal.
The package will stay closed after being resealed.

Figure 9.7 Peel-n-seal (by courtesy of Akerlund and Rausing, Lund, Sweden).

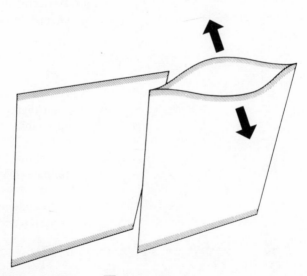

Easy-peel — Easy to open

The perfect answer to the opening of all flexible packages. Tremendously adaptable — can be used almost everywhere. Suitable when packaging a product from one web. Can easily be integrated with standard flexible laminate. Just pull to open.

Figure 9.8 Easy-peel (by courtesy of Akerlund and Rausing, Lund, Sweden).

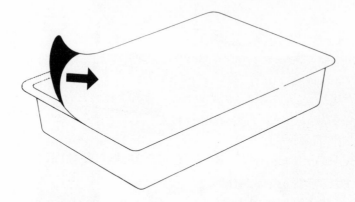

Transopeel — Easy to open

Makes it easy to open vacuum packs. Perfect when different top and bottom webs are being used.

Figure 9.9   Transopeel (by courtesy of Akerlund and Rausing, Lund, Sweden).

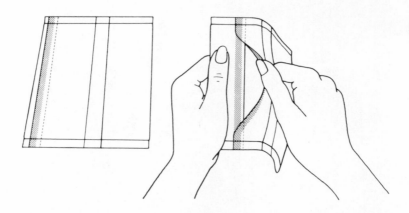

Securiseal — Easy to open and just as easy to reseal

A special sealing tape that is applied by the packaging machine during the packaging process. Makes all flexible packages easy to open and just as easy to reseal.

Figure 9.10   Securiseal (by courtesy of Akerlund and Rausing, Lund, Sweden).

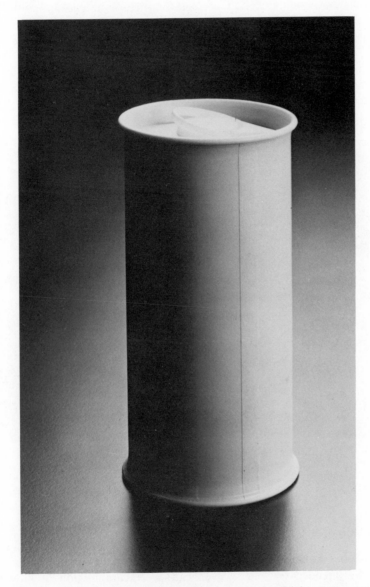

Figure 9.11 Originally a square/rectangular package, the new Let Pak has a more can-like appearance. The Let Pak is presented as the food can of tomorrow by its producer (Akerlund and Rausing, Lund, Sweden). An all-plastic container for sterilized food products.

Cekacan

The Cekacan (Figure 9.12) represents another food package with considerable future potential, made from laminated cartonboard. It is intended for dry products such as coffee, milk powder, and snacks, and can be made in various shapes—circular, oval, square or rectangular cross-section—and in sizes from 100 to 3000 ml. It is convenient to the consumer, as it is easy to open and dispense from this package. The producers claim it to be considerably cheaper than equivalent packages made from glass or metal.

Figure 9.12 The Cekacan for dry products offers great flexibility as to shape, size, opening facilities and laminate composition corresponding to product requirements. It also presents space savings of up to 97% compared to preproduced packagings. The Cekacan system is now available on a fully commercial scale in Britain, West Germany and the USA. (By courtesy of Akerlund and Rausing, Lund, Sweden.)

Microwave cooking

One of the most interesting current areas of development is in food products intended for microwave heating. A combination of food and packaging technologies allows the modern consumer to have available in the home food which can be taken from the freezer to the table in just a few minutes (Figure 9.13). Microwave heating requires not only a different package style but represents a totally new concept. These new products justify a higher price, because the modern consumer is willing to pay for convenience.

Figure 9.13 The traditional baby-food glass jar has become a convenient container for the growing microwave market. Shelf-stable, non-refrigerated heat-and-serve jars are now offered to consumers for soups, stews, spaghetti and sauce, etc.

The use of microwave ovens is considered to be one of the most important factors in the future development of food packaging. In the USA every other household has at least one microwave oven, and within 10–15 years it is expected that they will be as common in the home as TV sets are today. Similar but slower progress is expected in other sophisticated areas of the world. This will continue to lead to more and more foods being packaged in containers that will withstand microwave heating, with the result that the new co-extruded and laminated materials based on plastics and paperboard are taking over from traditional packages.

Microglas

Glass container manufacturers, together with the closure makers, White-cap, have responded with new ideas and market applications to the increasing use of microwave ovens (see above). One example is Microglas (Figure 9.14), a system for foods preserved in glass jars with a twist-off metal lid, presented in a carton suitable for microwave cooking. This is seen as a first-class innovation for the busy customer who is provided with a 'open-heat-and-eat' range of products which are shelf-stable and do not require refrigerated storage.

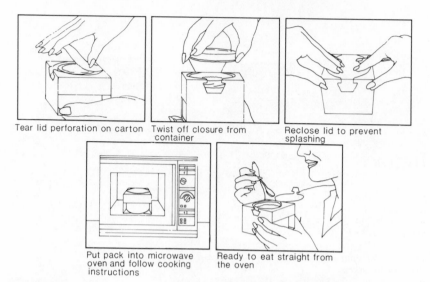

Tear lid perforation on carton   Twist off closure from   Reclose lid to prevent
                            container   splashing

Put pack into microwave   Ready to eat straight from
oven and follow cooking   the oven
instructions

Figure 9.14 The Microglas concept. The result of cooperation between the food producer, Rockware Glass Ltd, DRG Multiple Packaging Ltd and Whitecap International. Reproduced with permission from Rockware Glass Ltd.

Future developments

Packaging history shows very clearly that many product innovations have developed through packaging technological advances. Corporations who depend on constantly introducing new products are, however, understandably reluctant to initiate and fund developments involving a major technical breakthrough for very sound reasons. The budget for the development, testing, advertising and introductory market research of a single new product can be very high. New product introductions also have, on average, a high failure rate. Many food producing companies therefore rely on their packaging suppliers to carry out basic development work, and are reluctant to adopt new unproven packaging systems. The wise food manufacturer often contacts his packaging suppliers at the concept stage of a new product. Such involvement gives the package manufacturer far greater opportunities to suggest solutions to problems that can reduce the overall production costs, but still increase sales appeal and consumer convenience.

Index

physical distribution management 134,
    141, 142
physical properties of film 48, 49, 60
PKL 108
pollution 118
polyamide 48, 49
polyester 8, 14, 48, 49, 60
polyethylene 14, 48, 49, 60, 86, 91, 94,
    149
polypropylene 8, 14, 48, 49, 60
polystyrene 15, 48, 49, 60
potatoes 57
pouch burst test 12
pouch fabrication 8–10
poultry 43, 52
preservation methods 20, 37
product sterilization 25
Pure pak 28, 33, 87
PVAc 60
PVdC 16, 48, 49, 60
PVC 14, 48, 49, 116

quality control 50, 103

radiation 52 et seq.
rancidity 40
recycling 120–23
regulations 57–9
respiration rates of typical vegetables 42
retortable plastic packaging 1–19
retort pouch 8–13
retort processing 1–19, 22
retort pouch testing 11
Rigello 115, 116
Röntgen 52, 53

safety 40
Salmonella 55
salting food 1
sea foods 55, 56
scurvy 1
sealing 12, 32, 45, 88, 95
Serac 34
Servac 34
serve-from-the-tray 147
shelf life 36, 38, 42, 44, 56
    prediction of 62–85
Shriver, A.K. 1
Siderac system 28
smoking of foods 1, 20
snack foods 44, 158
socio-economic climate 147
spices 56
spoilage organisms 4

spoilage of food 4, 55
sprouting of potatoes 57
standards 59
Steriglen system 27
sterility calculations 23
sterilization, packaging 27, 56
sterilization, products 25, 57
Steritherm 25
storage 36
sun drying 1

testing 11, 50, 160
Tetra Brik 28, 33, 87–91, 151–4
Tetra Brik Aseptic 102, 151
thermal death time 5
thermal processing 4
thermal shock 3
thermoforms 16, 17
tie layers 16
tinplate 2
tomato ketchup 150
Torry Research Station 52
transmission rate 48, 49
Transofoil 154
transport 137, 138, 140–4
trays 14, 46
trippage 123
Tritello tub 149

UHT (ultra-high-temperature;
    ultra-heat-treatment) 32, 100
Ultralock 16
Ultramatic 25
Ultratherm 25
ultraviolet radiation 20, 27, 29, 30, 101
Uperizer 25
US Atomic Energy Commission 52, 53

vacuum packages 36, 47
vegetables 41, 42, 57
vitamin deficiency 1
volatiles 65
Votator 25

waste 124
water absorption 67
water vapour 63, 68, 70, 84
water vapour pressure 72
web sterilization 28
WVTR of packaging films 15, 49, 70

yoghurt 148

Z value 6